SHELSEAS MAGIC SQUARE
Magical Stars 2056

Magic Square is a mathematical formula containing rows, columns and diagonals with the same sum of integer as an entire whole. Over the years, around the world, people have enjoyed this magnificent symmetry of numbers.

Add each square of numbers in any row, column, or diagonal for the same sum of 2056.

256	255	254	253	5	6	7	8	9	10	11	12	244	243	242	241
240	239	238	20	21	22	23	233	232	26	27	28	29	227	226	225
224	223	222	221	37	38	39	40	41	42	43	44	212	211	210	209
208	50	206	205	204	54	55	56	57	58	59	197	196	195	63	193
65	66	67	189	188	187	186	72	73	183	182	181	180	78	79	80
81	82	83	84	172	171	170	169	168	167	166	165	93	94	95	96
97	98	99	100	156	155	154	153	152	151	150	149	109	110	111	112
113	143	115	116	117	139	138	137	136	135	134	124	125	126	130	128
129	127	131	132	133	123	122	121	120	119	118	140	141	142	114	144
145	146	147	148	108	107	106	105	104	103	102	101	157	158	159	160
161	162	163	164	92	91	90	89	88	87	86	85	173	174	175	176
177	178	179	77	76	75	74	184	185	71	70	69	68	190	191	192
64	194	62	61	60	198	199	200	201	202	203	53	52	51	207	49
48	47	46	45	213	214	215	216	217	218	219	220	36	35	34	33
32	31	30	228	229	230	231	25	24	234	235	236	237	19	18	17
16	15	14	13	245	246	247	248	249	250	251	252	4	3	2	1

2056

The structure of the number fill-ins [A thru D] will show you a creative method to sharpen your skills at counting & calculations to proof for a solved Magic Square.

SHELSEAS MAGIC SQUARE
Magical Stars 2056

Positive [C] Diagonals are shown for the same sum of 2056

256	255	254	253									244	243	242	241
240	239	238					233	232					227	226	225
224	223	222	221									212	211	210	209
208		206	205	204							197	196	195		193
			189	188	187	186			183	182	181	180			
				172	171	170	169	168	167	166	165				
				156	155	154	153	152	151	150	149				
	143				139	138	137	136	135	134				130	
	127				123	122	121	120	119	118				114	
				108	107	106	105	104	103	102	101				
				92	91	90	89	88	87	86	85				
			77	76	75	74			71	70	69	68			
64		62	61	60							53	52	51		49
48	47	46	45									36	35	34	33
32	31	30					25	24					19	18	17
16	15	14	13									4	3	2	1

Negative [C]

				5	6	7	8	9	10	11	12				
			20	21	22	23			26	27	28	29			
				37	38	39	40	41	42	43	44				
	50				54	55	56	57	58	59				63	
65	66	67					72	73					78	79	80
81	82	83	84									93	94	95	96
97	98	99	100									109	110	111	112
113		115	116	117							124	125	126		128
129		131	132	133							140	141	142		144
145	146	147	148									157	158	159	160
161	162	163	164									173	174	175	176
177	178	179					184	185					190	191	192
	194				198	199	200	201	202	203				207	
				213	214	215	216	217	218	219	220				
			228	229	230	231			234	235	236	237			
				245	246	247	248	249	250	251	252				

The mathematical formula $16 \times 16 + 1 \times 8 = 2056$ will introduce unique geometric patterns of shaded squares through the symmetric two parts in each Magic Square.

SHELSEAS MAGIC SQUARE
Magical Stars 2056

Description

Magical Stars 2056 will offer [20] Magic Squares.

Integer: In each Magic Square the sum of 2056 as an entire whole is symbolic to the mathematical formula 16 X 16 + 1 X 8 = 2056.

Double Even: Even number of rows & columns, divide by two, for even number of shaded squares in each row & column.

Symmetric: Two parts of Positive & Negative in each Magic Square.

Geometric: Unique patterns of shaded squares in each Magic Square.

Guideline: The mathematical formula 4 X 4 + 1 X 2 = 34 will be shown as a solved Magic Square in the number fill-ins [A thru D] format.

Extension: The mathematical formula 16 X 16 + 1 X 8 = 2056 will be shown as an extension in the number fill-ins [A thru D] in each Magic Square.

Objective: Combine both Positive & Negative number fill-ins [A thru D] in each Magic Square. Add each square of numbers in any row, column, or diagonal to proof the same sum of 2056 for a solved Magic Square.

"Welcome to my world of Magic Squares"

Created by,
Peggy Brown

Guideline

A

Positive [A] Start with the top row, left to right in each row, going down. Begin with the number shown hint, count each square, only fill-in shaded squares.

Negative [A] Start with the bottom row, right to left in each row, going up. Begin with the number shown hint, count each square, only fill-in shaded squares.

B

Positive [B] Start with the top row, right to left in each row, going down. Begin with the number shown hint, count each square, only fill-in shaded squares.

Negative [B] Start with the bottom row, left to right in each row, going up. Begin with the number shown hint, count each square, only fill-in shaded squares.

C

Positive [C] Start with the bottom row, right to left in each row, going up. Begin with the number shown hint, count each square, only fill-in shaded squares.

Negative [C] Start with the top row, left to right in each row, going down. Begin with the number shown hint, count each square, only fill-in shaded squares.

D

Positive [D] Start with the bottom row, left to right in each row, going up. Begin with the number shown hint, count each square, only fill-in shaded squares.

Negative [D] Start with the top row, right to left in each row, going down. Begin with the number shown hint, count each square, only fill-in shaded squares.

SHELSEAS MAGIC SQUARE
Magical Stars 2056

The Number Fill-ins [A] Will Be Your Guideline
For A Solved Double Even – Magic Square.

Number
Fill-ins

[A]

Positive [A] Start with the top row, left to right in each row, going down. Begin with the number shown hint, count each square, only fill-in shaded squares.

Negative [A] Start with the bottom row, right to left in each row, going up. Begin with the number shown hint, count each square, only fill-in shaded squares.

Objective: Combine both Positive & Negative number fill-ins [A] in each Magic Square. Add each square of numbers in any row, column, or diagonal to proof the same sum of 2056 for a solved Magic Square.

Positive [A] Start with the top row, left to right in each row, going down. Begin with the number 1, count each square, only fill-in shaded squares.

1			4
	6	7	
	10	11	
13			16

Negative [A] Start with the bottom row, right to left in each row, going up. Begin with the number 2, count each square, only fill-in shaded squares.

	15	14	
12			9
8			5
	3	2	

Combination of both Positive & Negative number fill-ins [A] is shown for a solved Magic Square. Objective: Add each square of numbers in any row, column, or diagonal to proof the same sum of 34.

1	15	14	4
12	6	7	9
8	10	11	5
13	3	2	16

Magic Square
Number Fill-ins [A]
Mathematical Formula 4 X 4 + 1 X 2 = 34

SHELSEAS MAGIC SQUARE
Magical Stars 2056

Positive [A] Start with the top row, left to right in each row, going down. Begin with the number 1, count each square, only fill-in shaded squares. Diagonals are shown for the same sum of 2056. Helpful hints below.

1															16
	18													31	
		35											46		
			52									61			
				69							76				
					86					91					
						103			106						
							120	121							
							136	137							
						151			154						
					166					171					
				181							188				
			196									205			
		211											222		
	226													239	
241															256

Negative [A] Start with the bottom row, right to left in each row, going up. Begin with the number 5, count each square, only fill-in shaded squares. Helpful hints below.

				252							245				
		238											227		
	223													210	
						201	200								
192															177
					170				167						
				155						150					
			141									132			
			125									116			
				107						102					
					90				87						
80															65
						57	56								
	47													34	
		30											19		
				12							5				

Magic Square - Number Fill-ins [A] [1] - Mathematical Formula 16 X 16 + 1 X 8 = 2056

SHELSEAS MAGIC SQUARE
Magical Stars 2056

Positive [A] Start with the top row, left to right in each row, going down. Begin with the number 1, count each square, only fill-in same shaded squares. Diagonals are shown for the same sum of 2056.

Negative [A] Start with the bottom row, right to left in each row, going up. Begin with the number 5, count each square, only fill-in same shaded squares.

Objective: Combine both Positive & Negative number fill-ins [A] [1]. Add each square of numbers in any row, column, or diagonal to proof the same sum of 2056 for a solved Magic Square. Helpful hints below.

1				252							245				16
	18	238											227	31	
	223	35											46	210	
			52			201	200				61				
192				69							76				177
					86	170		167	91						
					155	103		106	150						
			141			120	121				132				
			125			136	137				116				
					107	151		154	102						
					166	90		87	171						
80				181							188				65
			196			57	56				205				
	47	211											222	34	
	226	30											19	239	
241				12							5				256

Magic Square
Number Fill-ins [A] [1]
Mathematical Formula 16 X 16 + 1 X 8 = 2056

SHELSEAS MAGIC SQUARE
Magical Stars 2056

Positive [A] Start with the top row, left to right in each row, going down. Begin with the number 1, count each square, only fill-in shaded squares. Diagonals are shown for the same sum of 2056. Helpful hints below.

1															16
	18													31	
		35											46		
			52									61			
				69							76				
					86					91					
						103			106						
							120	121							
							136	137							
						151			154						
					166					171					
				181							188				
			196									205			
		211											222		
	226													239	
241															256

Negative [A] Start with the bottom row, right to left in each row, going up. Begin with the number 3, count each square, only fill-in shaded squares. Helpful hints below.

		254											243		
			236									229			
224															209
						201	200								
	191													178	
					170			167							
				155					150						
			141									132			
			125									116			
				107					102						
					90			87							
	79													66	
						57	56								
48															33
			28									21			
		14											3		

Magic Square - Number Fill-ins [A] [2] - Mathematical Formula 16 X 16 + 1 X 8 = 2056

SHELSEAS MAGIC SQUARE
Magical Stars 2056

Positive [A] Start with the top row, left to right in each row, going down. Begin with the number 1, count each square, only fill-in same shaded squares. Diagonals are shown for the same sum of 2056.

Negative [A] Start with the bottom row, right to left in each row, going up. Begin with the number 3, count each square, only fill-in same shaded squares.

Objective: Combine both Positive & Negative number fill-ins [A] [2]. Add each square of numbers in any row, column, or diagonal to proof the same sum of 2056 for a solved Magic Square. Helpful hints below.

1		254											243		16
	18		236								229		31		
224	35												46		209
		52				201	200					61			
	191		69								76		178		
				86	170			167	91						
				155	103			106	150						
			141			120	121				132				
			125			136	137				116				
				107	151			154	102						
				166	90			87	171						
	79			181							188		66		
			196			57	56				205				
48		211											222		33
	226		28								21		239		
241		14											3		256

Magic Square
Number Fill-ins [A] [2]
Mathematical Formula 16 X 16 + 1 X 8 = 2056

SHELSEAS MAGIC SQUARE
Magical Stars 2056

Positive [A] Start with the top row, left to right in each row, going down. Begin with the number 1, count each square, only fill-in shaded squares. Diagonals are shown for the same sum of 2056. Helpful hints below.

1															16
	18													31	
		35											46		
			52									61			
				69							76				
					86					91					
						103			106						
							120	121							
							136	137							
						151			154						
					166					171					
				181							188				
			196									205			
		211											222		
	226													239	
241															256

Negative [A] Start with the bottom row, right to left in each row, going up. Begin with the number 3, count each square, only fill-in shaded squares. Helpful hints below.

		254											243		
						233	232								
					218			215							
				203					198						
	191													178	
176															161
				156						149					
			141									132			
			125									116			
				108						101					
96															81
	79													66	
				59						54					
					42			39							
						25	24								
		14											3		

Magic Square - Number Fill-ins [A] [3] - Mathematical Formula 16 X 16 + 1 X 8 = 2056

SHELSEAS MAGIC SQUARE
Magical Stars 2056

Positive [A] Start with the top row, left to right in each row, going down. Begin with the number 1, count each square, only fill-in same shaded squares. Diagonals are shown for the same sum of 2056.

Negative [A] Start with the bottom row, right to left in each row, going up. Begin with the number 3, count each square, only fill-in same shaded squares.

Objective: Combine both Positive & Negative number fill-ins [A] [3]. Add each square of numbers in any row, column, or diagonal to proof the same sum of 2056 for a solved Magic Square. Helpful hints below.

1	2	3	4	5	6	7	8	9	10	11	12	13	14	15	16
1		254												243	16
	18					233	232							31	
		35			218			215					46		
			52		203					198		61			
	191			69							76			178	
176					86					91					161
				156		103			106		149				
			141				120	121				132			
			125				136	137				116			
				108		151			154		101				
96					166					171					81
	79			181							188			66	
			196		59					54		205			
		211				42			39					222	
	226						25	24						239	
241		14												3	256

Magic Square
Number Fill-ins [A] [3]
Mathematical Formula 16 X 16 + 1 X 8 = 2056

SHELSEAS MAGIC SQUARE
Magical Stars 2056

Positive [A] Start with the top row, left to right in each row, going down. Begin with the number 1, count each square, only fill-in shaded squares. Diagonals are shown for the same sum of 2056. Helpful hints below.

1															16
	18													31	
		35											46		
			52									61			
				69							76				
					86					91					
						103			106						
							120	121							
							136	137							
						151			154						
					166					171					
				181							188				
			196									205			
		211											222		
	226													239	
241															256

Negative [A] Start with the bottom row, right to left in each row, going up. Begin with the number 3, count each square, only fill-in shaded squares. Helpful hints below.

		254											243		
			237									228			
224															209
					203					198					
	191													178	
							169	168							
				156							149				
						138			135						
						122			119						
				108							101				
							89	88							
	79													66	
					59					54					
48															33
			29									20			
		14											3		

Magic Square - Number Fill-ins [A] [4] - Mathematical Formula 16 X 16 + 1 X 8 = 2056

SHELSEAS MAGIC SQUARE
Magical Stars 2056

Positive [A] Start with the top row, left to right in each row, going down. Begin with the number 1, count each square, only fill-in same shaded squares. Diagonals are shown for the same sum of 2056.

Negative [A] Start with the bottom row, right to left in each row, going up. Begin with the number 3, count each square, only fill-in same shaded squares.

Objective: Combine both Positive & Negative number fill-ins [A] [4]. Add each square of numbers in any row, column, or diagonal to proof the same sum of 2056 for a solved Magic Square. Helpful hints below.

1	2	3	4	5	6	7	8	9	10	11	12	13	14	15	16
1		254											243		16
	18		237									228		31	
224		35											46		209
			52		203					198		61			
	191			69						76				178	
				86		169	168		91						
			156		103			106		149					
					138	120	121	135							
					122	136	137	119							
			108		151			154		101					
				166		89	88		171						
	79			181						188				66	
			196		59					54		205			
48		211											222		33
	226		29									20		239	
241		14											3		256

Magic Square
Number Fill-ins [A] [4]
Mathematical Formula $16 \times 16 + 1 \times 8 = 2056$

SHELSEAS MAGIC SQUARE
Magical Stars 2056

Positive [A] Start with the top row, left to right in each row, going down. Begin with the number 1, count each square, only fill-in shaded squares. Diagonals are shown for the same sum of 2056. Helpful hints below.

1															16
	18													31	
		35											46		
			52									61			
				69							76				
					86					91					
						103			106						
							120	121							
							136	137							
						151			154						
					166					171					
				181							188				
			196									205			
		211											222		
	226													239	
241															256

Negative [A] Start with the bottom row, right to left in each row, going up. Begin with the number 4, count each square, only fill-in shaded squares. Helpful hints below.

			253									244			
				236							229				
						218			215						
208															193
	191													178	
							169	168							
		158											147		
					139					134					
					123					118					
		110											99		
							89	88							
	79													66	
64															49
						42			39						
				28							21				
			13									4			

Magic Square - Number Fill-ins [A] [5] - Mathematical Formula 16 X 16 + 1 X 8 = 2056

SHELSEAS MAGIC SQUARE
Magical Stars 2056

Positive [A] Start with the top row, left to right in each row, going down. Begin with the number 1, count each square, only fill-in same shaded squares. Diagonals are shown for the same sum of 2056.

Negative [A] Start with the bottom row, right to left in each row, going up. Begin with the number 3, count each square, only fill-in same shaded squares.

Objective: Combine both Positive & Negative number fill-ins [A] [5]. Add each square of numbers in any row, column, or diagonal to proof the same sum of 2056 for a solved Magic Square. Helpful hints below.

1	2	3	4	5	6	7	8	9	10	11	12	13	14	15	16
1			253									244			16
	18			236							229			31	
		35				218			215				46		
208			52									61			193
	191			69							76			178	
					86		169	168		91					
		158				103			106				147		
					139		120	121		134					
					123		136	137		118					
		110				151			154				99		
					166		89	88		171					
	79			181							188			66	
64			196									205			49
		211				42			39				222		
	226			28							21			239	
241			13									4			256

Magic Square
Number Fill-ins [A] [5]
Mathematical Formula 16 X 16 + 1 X 8 = 2056

SHELSEAS MAGIC SQUARE
Magical Stars 2056

Solved

[A]

Magic Square
Number Fill-ins [A]
Mathematical Formula 16 X 16 + 1 X 8 = 2056

SHELSEAS MAGIC SQUARE
Magical Stars 2056

Positive [A] Diagonals are shown for the same sum of 2056

1	2	3	4									13	14	15	16
17	18			21			24	25			28			31	32
33		35			38	39			42	43			46		48
49			52		54	55			58	59		61			64
	66			69		71	72	73	74		76			79	
		83	84		86		88	89		91		93	94		
		99	100	101		103			106		108	109	110		
	114			117	118		120	121		123	124			127	
	130			133	134		136	137		139	140			143	
		147	148	149		151			154		156	157	158		
		163	164		166		168	169		171		173	174		
	178			181		183	184	185	186		188			191	
193			196		198	199			202	203		205			208
209		211			214	215			218	219			222		224
225	226			229			232	233			236			239	240
241	242	243	244									253	254	255	256

Negative [A]

				252	251	250	249	248	247	246	245				
		238	237		235	234			231	230		228	227		
	223		221	220			217	216			213	212		210	
	207	206		204			201	200			197		195	194	
192		190	189		187					182		180	179		177
176	175			172		170			167		165			162	161
160	159				155		153	152		150				146	145
144		142	141			138			135			132	131		129
128		126	125			122			119			116	115		113
112	111				107		105	104		102				98	97
96	95			92		90			87		85			82	81
80		78	77		75					70		68	67		65
	63	62		60			57	56			53		51	50	
	47		45	44			41	40			37	36		34	
		30	29		27	26			23	22		20	19		
				12	11	10	9	8	7	6	5				

Magic Square
Number Fill-ins [A] [1]
Mathematical Formula 16 X 16 + 1 X 8 = 2056

SHELSEAS MAGIC SQUARE
Magical Stars 2056

Combination of both Positive & Negative number fill-ins [A] [1] is shown for a solved Magic Square. Objective: Add each square of numbers in any row, column, or diagonal to proof the same sum of 2056.

1	2	3	4	252	251	250	249	248	247	246	245	13	14	15	16
17	18	238	237	21	235	234	24	25	231	230	28	228	227	31	32
33	223	35	221	220	38	39	217	216	42	43	213	212	46	210	48
49	207	206	52	204	54	55	201	200	58	59	197	61	195	194	64
192	66	190	189	69	187	71	72	73	74	182	76	180	179	79	177
176	175	83	84	172	86	170	88	89	167	91	165	93	94	162	161
160	159	99	100	101	155	103	153	152	106	150	108	109	110	146	145
144	114	142	141	117	118	138	120	121	135	123	124	132	131	127	129
128	130	126	125	133	134	122	136	137	119	139	140	116	115	143	113
112	111	147	148	149	107	151	105	104	154	102	156	157	158	98	97
96	95	163	164	92	166	90	168	169	87	171	85	173	174	82	81
80	178	78	77	181	75	183	184	185	186	70	188	68	67	191	65
193	63	62	196	60	198	199	57	56	202	203	53	205	51	50	208
209	47	211	45	44	214	215	41	40	218	219	37	36	222	34	224
225	226	30	29	229	27	26	232	233	23	22	236	20	19	239	240
241	242	243	244	12	11	10	9	8	7	6	5	253	254	255	256

2056

Magic Square
Number Fill-ins [A] [1]
Mathematical Formula 16 X 16 + 1 X 8 = 2056

SHELSEAS MAGIC SQUARE
Magical Stars 2056

Positive [A] Diagonals are shown for the same sum of 2056

1	2		4			7			10			13		15	16
17	18	19			22					27			30	31	32
	34	35	36			39			42			45	46	47	
49		51	52	53							60	61	62		64
			68	69	70		72	73		75	76	77			
	82			85	86		88	89		91	92			95	
97		99				103	104	105	106				110		112
				117	118	119	120	121	122	123	124				
				133	134	135	136	137	138	139	140				
145		147				151	152	153	154				158		160
	162			165	166		168	169		171	172			175	
			180	181	182		184	185		187	188	189			
193		195	196	197							204	205	206		208
	210	211	212			215			218			221	222	223	
225	226	227			230					235			238	239	240
241	242		244			247			250			253		255	256

Negative [A]

		254		252	251		249	248		246	245		243		
			237	236		234	233	232	231		229	228			
224				220	219		217	216		214	213				209
	207				203	202	201	200	199	198				194	
192	191	190				186			183				179	178	177
176		174	173			170			167			164	163		161
	159		157	156	155					150	149	148		146	
144	143	142	141									132	131	130	129
128	127	126	125									116	115	114	113
	111		109	108	107					102	101	100		98	
96		94	93			90			87			84	83		81
80	79	78				74			71				67	66	65
	63				59	58	57	56	55	54				50	
48				44	43		41	40		38	37				33
			29	28		26	25	24	23		21	20			
		14		12	11		9	8		6	5		3		

Magic Square
Number Fill-ins [A] [2]
Mathematical Formula 16 X 16 + 1 X 8 = 2056

SHELSEAS MAGIC SQUARE
Magical Stars 2056

Combination of both Positive & Negative number fill-ins [A] [2] is shown for a solved Magic Square. Objective: Add each square of numbers in any row, column, or diagonal to proof the same sum of 2056.

1	2	254	4	252	251	7	249	248	10	246	245	13	243	15	16
17	18	19	237	236	22	234	233	232	231	27	229	228	30	31	32
224	34	35	36	220	219	39	217	216	42	214	213	45	46	47	209
49	207	51	52	53	203	202	201	200	199	198	60	61	62	194	64
192	191	190	68	69	70	186	72	73	183	75	76	77	179	178	177
176	82	174	173	85	86	170	88	89	167	91	92	164	163	95	161
97	159	99	157	156	155	103	104	105	106	150	149	148	110	146	112
144	143	142	141	117	118	119	120	121	122	123	124	132	131	130	129
128	127	126	125	133	134	135	136	137	138	139	140	116	115	114	113
145	111	147	109	108	107	151	152	153	154	102	101	100	158	98	160
96	162	94	93	165	166	90	168	169	87	171	172	84	83	175	81
80	79	78	180	181	182	74	184	185	71	187	188	189	67	66	65
193	63	195	196	197	59	58	57	56	55	54	204	205	206	50	208
48	210	211	212	44	43	215	41	40	218	38	37	221	222	223	33
225	226	227	29	28	230	26	25	24	23	235	21	20	238	239	240
241	242	14	244	12	11	247	9	8	250	6	5	253	3	255	256

2056

Magic Square
Number Fill-ins [A] [2]
Mathematical Formula 16 X 16 + 1 X 8 = 2056

SHELSEAS MAGIC SQUARE
Magical Stars 2056

Positive [A] Diagonals are shown for the same sum of 2056

1	2					7	8	9	10					15	16
	18	19			22	23			26	27			30	31	
		35	36	37	38				43	44	45	46			
49			52	53		56	57			60	61				64
65			68	69		72	73			76	77				80
		83	84	85	86					91	92	93	94		
	98	99			102	103			106	107			110	111	
113	114					119	120	121	122					127	128
129	130					135	136	137	138					143	144
	146	147			150	151			154	155			158	159	
		163	164	165	166					171	172	173	174		
177			180	181			184	185			188	189			192
193			196	197			200	201			204	205			208
		211	212	213	214					219	220	221	222		
	226	227			230	231			234	235			238	239	
241	242					247	248	249	250					255	256

Negative [A]

		254	253	252	251					246	245	244	243		
240			237	236			233	232			229	228			225
224	223					218	217	216	215					210	209
	207	206			203	202			199	198			195	194	
	191	190			187	186			183	182			179	178	
176	175					170	169	168	167					162	161
160			157	156			153	152			149	148			145
		142	141	140	139					134	133	132	131		
		126	125	124	123					118	117	116	115		
112			109	108			105	104			101	100			97
96	95					90	89	88	87					82	81
	79	78			75	74			71	70			67	66	
	63	62			59	58			55	54			51	50	
48	47					42	41	40	39					34	33
32			29	28			25	24			21	20			17
		14	13	12	11					6	5	4	3		

Magic Square
Number Fill-ins [A] [3]
Mathematical Formula 16 X 16 + 1 X 8 = 2056

SHELSEAS MAGIC SQUARE
Magical Stars 2056

Combination of both Positive & Negative number fill-ins [A] [3] is shown for a solved Magic Square. Objective: Add each square of numbers in any row, column, or diagonal to proof the same sum of 2056.

1	2	254	253	252	251	7	8	9	10	246	245	244	243	15	16
240	18	19	237	236	22	23	233	232	26	27	229	228	30	31	225
224	223	35	36	37	38	218	217	216	215	43	44	45	46	210	209
49	207	206	52	53	203	202	56	57	199	198	60	61	195	194	64
65	191	190	68	69	187	186	72	73	183	182	76	77	179	178	80
176	175	83	84	85	86	170	169	168	167	91	92	93	94	162	161
160	98	99	157	156	102	103	153	152	106	107	149	148	110	111	145
113	114	142	141	140	139	119	120	121	122	134	133	132	131	127	128
129	130	126	125	124	123	135	136	137	138	118	117	116	115	143	144
112	146	147	109	108	150	151	105	104	154	155	101	100	158	159	97
96	95	163	164	165	166	90	89	88	87	171	172	173	174	82	81
177	79	78	180	181	75	74	184	185	71	70	188	189	67	66	192
193	63	62	196	197	59	58	200	201	55	54	204	205	51	50	208
48	47	211	212	213	214	42	41	40	39	219	220	221	222	34	33
32	226	227	29	28	230	231	25	24	234	235	21	20	238	239	17
241	242	14	13	12	11	247	248	249	250	6	5	4	3	255	256

2056

Magic Square
Number Fill-ins [A] [3]
Mathematical Formula 16 X 16 + 1 X 8 = 2056

SHELSEAS MAGIC SQUARE
Magical Stars 2056

Positive [A] Diagonals are shown for the same sum of 2056

1	2					7	8	9	10					15	16
17	18	19					24	25					30	31	32
	34	35	36	37						44	45	46	47		
	50	51	52	53						60	61	62	63		
			68	69	70	71			74	75	76	77			
			84	85	86	87			90	91	92	93			
97				102	103	104	105	106	107						112
113		115			118		120	121		123			126		128
129		131			134		136	137		139			142		144
145					150	151	152	153	154	155					160
			164	165	166	167			170	171	172	173			
			180	181	182	183			186	187	188	189			
	194	195	196	197						204	205	206	207		
	210	211	212	213						220	221	222	223		
225	226	227					232	233					238	239	240
241	242					247	248	249	250					255	256

Negative [A]

		254	253	252	251					246	245	244	243		
		237	236	235	234				231	230	229	228			
224				219	218	217	216	215	214						209
208				203	202	201	200	199	198						193
192	191	190				185	184						179	178	177
176	175	174				169	168						163	162	161
	159	158	157	156						149	148	147	146		
	143		141	140		138			135	133	132		130		
	127		125	124		122			119	117	116		114		
	111	110	109	108						101	100	99	98		
96	95	94				89	88						83	82	81
80	79	78				73	72						67	66	65
64				59	58	57	56	55	54						49
48				43	42	41	40	39	38						33
		29	28	27	26				23	22	21	20			
		14	13	12	11				6	5	4	3			

Magic Square
Number Fill-ins [A] [4]
Mathematical Formula 16 X 16 + 1 X 8 = 2056

SHELSEAS MAGIC SQUARE
Magical Stars 2056

Combination of both Positive & Negative number fill-ins [A] [4] is shown for a solved
Magic Square. Objective: Add each square of numbers in any row, column, or diagonal
to proof the same sum of 2056.

1	2	254	253	252	251	7	8	9	10	246	245	244	243	15	16
17	18	19	237	236	235	234	24	25	231	230	229	228	30	31	32
224	34	35	36	37	219	218	217	216	215	214	44	45	46	47	209
208	50	51	52	53	203	202	201	200	199	198	60	61	62	63	193
192	191	190	68	69	70	71	185	184	74	75	76	77	179	178	177
176	175	174	84	85	86	87	169	168	90	91	92	93	163	162	161
97	159	158	157	156	102	103	104	105	106	107	149	148	147	146	112
113	143	115	141	140	118	138	120	121	135	123	133	132	126	130	128
129	127	131	125	124	134	122	136	137	119	139	117	116	142	114	144
145	111	110	109	108	150	151	152	153	154	155	101	100	99	98	160
96	95	94	164	165	166	167	89	88	170	171	172	173	83	82	81
80	79	78	180	181	182	183	73	72	186	187	188	189	67	66	65
64	194	195	196	197	59	58	57	56	55	54	204	205	206	207	49
48	210	211	212	213	43	42	41	40	39	38	220	221	222	223	33
225	226	227	29	28	27	26	232	233	23	22	21	20	238	239	240
241	242	14	13	12	11	247	248	249	250	6	5	4	3	255	256

2056

Magic Square
Number Fill-ins [A] [4]
Mathematical Formula 16 X 16 + 1 X 8 = 2056

SHELSEAS MAGIC SQUARE
Magical Stars 2056

Positive [A] Diagonals are shown for the same sum of 2056

1	2	3					8	9					14	15	16
17	18		20			23			26			29		31	32
33		35	36		38					43		45	46		48
	50	51	52				56	57				61	62	63	
				69	70	71	72	73	74	75	76				
		83		85	86	87			90	91	92		94		
	98			101	102	103			106	107	108			111	
113			116	117			120	121			124	125			128
129			132	133			136	137			140	141			144
	146			149	150	151			154	155	156			159	
		163		165	166	167			170	171	172		174		
				181	182	183	184	185	186	187	188				
	194	195	196				200	201				205	206	207	
209		211	212		214					219		221	222		224
225	226		228			231			234			237		239	240
241	242	243					248	249					254	255	256

Negative [A]

			253	252	251	250			247	246	245	244			
		238		236	235		233	232		230	229		227		
	223			220		218	217	216	215		213			210	
208				204	203	202			199	198	197				193
192	191	190	189									180	179	178	177
176	175		173				169	168				164		162	161
160		158	157				153	152				148	147		145
	143	142			139	138			135	134			131	130	
	127	126			123	122			119	118			115	114	
112		110	109				105	104				100	99		97
96	95		93				89	88				84		82	81
80	79	78	77									68	67	66	65
64				60	59	58			55	54	53				49
	47			44		42	41	40	39		37			34	
		30		28	27		25	24		22	21		19		
			13	12	11	10			7	6	5	4			

Magic Square
Number Fill-ins [A] [5]
Mathematical Formula 16 X 16 + 1 X 8 = 2056

SHELSEAS MAGIC SQUARE
Magical Stars 2056

Combination of both Positive & Negative number fill-ins [A] [5] is shown for a solved
Magic Square. Objective: Add each square of numbers in any row, column, or diagonal
to proof the same sum of 2056.

1	2	3	253	252	251	250	8	9	247	246	245	244	14	15	16
17	18	238	20	236	235	23	233	232	26	230	229	29	227	31	32
33	223	35	36	220	38	218	217	216	215	43	213	45	46	210	48
208	50	51	52	204	203	202	56	57	199	198	197	61	62	63	193
192	191	190	189	69	70	71	72	73	74	75	76	180	179	178	177
176	175	83	173	85	86	87	169	168	90	91	92	164	94	162	161
160	98	158	157	101	102	103	153	152	106	107	108	148	147	111	145
113	143	142	116	117	139	138	120	121	135	134	124	125	131	130	128
129	127	126	132	133	123	122	136	137	119	118	140	141	115	114	144
112	146	110	109	149	150	151	105	104	154	155	156	100	99	159	97
96	95	163	93	165	166	167	89	88	170	171	172	84	174	82	81
80	79	78	77	181	182	183	184	185	186	187	188	68	67	66	65
64	194	195	196	60	59	58	200	201	55	54	53	205	206	207	49
209	47	211	212	44	214	42	41	40	39	219	37	221	222	34	224
225	226	30	228	28	27	231	25	24	234	22	21	237	19	239	240
241	242	243	13	12	11	10	248	249	7	6	5	4	254	255	256

2056

Magic Square
Number Fill-ins [A] [5]
Mathematical Formula 16 X 16 + 1 X 8 = 2056

The Number Fill-ins [B] Will Be Your Guideline
For A Solved Double Even – Magic Square.

Number
Fill-ins

[B]

Positive [B] Start with the top row, right to left in each row, going down. Begin with the number shown hint, count each square, only fill-in shaded squares.

Negative [B] Start with the bottom row, left to right in each row, going up. Begin with the number shown hint, count each square, only fill-in shaded squares.

Objective: Combine both Positive & Negative number fill-ins [B] in each Magic Square. Add each square of numbers in any row, column, or diagonal to proof the same sum of 2056 for a solved Magic Square.

Positive [B] Start with the top row, right to left in each row, going down. Begin with the number 1, count each square, only fill-in shaded squares.

4			1
	7	6	
	11	10	
16			13

Negative [B] Start with the bottom row, left to right in each row, going up. Begin with the number 2, count each square, only fill-in shaded squares.

	14	15	
9			12
5			8
	2	3	

Combination of both Positive & Negative number fill-ins [B] is shown for a solved Magic Square. Objective: Add each square of numbers in any row, column, or diagonal to proof the same sum of 34.

4	14	15	1
9	7	6	12
5	11	10	8
16	2	3	13

Magic Square
Number Fill-ins [B]
Mathematical Formula 4 X 4 + 1 X 2 = 34

SHELSEAS MAGIC SQUARE
Magical Stars 2056

Positive [B] Start with the top row, right to left in each row, going down. Begin with the number 1, count each square, only fill-in shaded squares. Diagonals are shown for the same sum of 2056. Helpful hints below.

16															1
	31													18	
		46											35		
			61									52			
				76							69				
					91					86					
						106			103						
							121	120							
							137	136							
						154			151						
					171					166					
				188							181				
			205									196			
		222											211		
	239													226	
256															241

Negative [B] Start with the bottom row, left to right in each row, going up. Begin with the number 5, count each square, only fill-in shaded squares. Helpful hints below.

				245							252				
		227											238		
	210													223	
							200	201							
177															192
						167			170						
					150					155					
				132							141				
				116							125				
					102					107					
						87			90						
65															80
							56	57							
	34													47	
		19											30		
				5							12				

Magic Square - Number Fill-ins [B] [1] - Mathematical Formula 16 X 16 + 1 X 8 = 2056

SHELSEAS MAGIC SQUARE
Magical Stars 2056

Positive [B] Start with the top row, right to left in each row, going down. Begin with the number 1, count each square, only fill-in same shaded squares. Diagonals are shown for the same sum of 2056.

Negative [B] Start with the bottom row, left to right in each row, going up. Begin with the number 5, count each square, only fill-in same shaded squares.

Objective: Combine both Positive & Negative number fill-ins [B] [1]. Add each square of numbers in any row, column, or diagonal to proof the same sum of 2056 for a solved Magic Square. Helpful hints below.

16				245						252					1
	31	227										238	18		
	210	46										35	223		
			61			200	201				52				
177				76							69				192
					91	167			170	86					
					150	106			103	155					
			132				121	120				141			
			116				137	136				125			
					102	154			151	107					
					171	87			90	166					
65				188							181				80
			205			56	57				196				
	34	222										211	47		
	239	19										30	226		
256				5						12					241

Magic Square
Number Fill-ins [B] [1]
Mathematical Formula 16 X 16 + 1 X 8 = 2056

SHELSEAS MAGIC SQUARE
Magical Stars 2056

Positive [B] Start with the top row, right to left in each row, going down. Begin with the number 1, count each square, only fill-in shaded squares. Diagonals are shown for the same sum of 2056. Helpful hints below.

16															1
	31													18	
		46											35		
			61									52			
				76							69				
					91					86					
						106			103						
							121	120							
							137	136							
						154			151						
					171					166					
				188							181				
			205									196			
		222											211		
	239													226	
256															241

Negative [B] Start with the bottom row, left to right in each row, going up. Begin with the number 3, count each square, only fill-in shaded squares. Helpful hints below.

		243												254	
						232	233								
209															224
				198							203				
					183					186					
			164									173			
				149							156				
	130													143	
	114													127	
				101							108				
			84									93			
					71					74					
				54							59				
33															48
						24	25								
		3												14	

Magic Square - Number Fill-ins [B] [2] - Mathematical Formula 16 X 16 + 1 X 8 = 2056

SHELSEAS MAGIC SQUARE
Magical Stars 2056

Positive [B] Start with the top row, right to left in each row, going down. Begin with the number 1, count each square, only fill-in same shaded squares. Diagonals are shown for the same sum of 2056.

Negative [B] Start with the bottom row, left to right in each row, going up. Begin with the number 3, count each square, only fill-in same shaded squares.

Objective: Combine both Positive & Negative number fill-ins [B] [2]. Add each square of numbers in any row, column, or diagonal to proof the same sum of 2056 for a solved Magic Square. Helpful hints below.

16		243											254		1
	31						232	233						18	
209		46											35		224
			61		198					203		52			
				76		183			186		69				
			164		91					86		173			
				149		106			103		156				
	130						121	120						143	
	114						137	136						127	
				101		154			151		108				
			84		171					166		93			
				188		71			74		181				
			205		54					59		196			
33		222											211		48
	239						24	25						226	
256		3											14		241

Magic Square
Number Fill-ins [B] [2]
Mathematical Formula 16 X 16 + 1 X 8 = 2056

SHELSEAS MAGIC SQUARE
Magical Stars 2056

Positive [B] Start with the top row, right to left in each row, going down. Begin with the number 1, count each square, only fill-in shaded squares. Diagonals are shown for the same sum of 2056. Helpful hints below.

16															1
	31													18	
		46											35		
			61									52			
				76							69				
					91					86					
						106			103						
							121	120							
							137	136							
						154			151						
					171					166					
				188							181				
			205									196			
		222											211		
	239													226	
256															241

Negative [B] Start with the bottom row, left to right in each row, going up. Begin with the number 5, count each square, only fill-in shaded squares. Helpful hints below.

				245							252				
							232	233							
209															224
						199			202						
			180									189			
	162													175	
					150					155					
		131											142		
		115											126		
					102					107					
	82													95	
			68									77			
					55				58						
33															48
							24	25							
				5							12				

Magic Square - Number Fill-ins [B] [3] - Mathematical Formula 16 X 16 + 1 X 8 = 2056

SHELSEAS MAGIC SQUARE
Magical Stars 2056

Positive [B] Start with the top row, right to left in each row, going down. Begin with the number 1, count each square, only fill-in same shaded squares. Diagonals are shown for the same sum of 2056.

Negative [B] Start with the bottom row, left to right in each row, going up. Begin with the number 5, count each square, only fill-in same shaded squares.

Objective: Combine both Positive & Negative number fill-ins [B] [3]. Add each square of numbers in any row, column or diagonal to proof the same sum of 2056 for a solved Magic Square. Helpful hints below.

16				245							252				1
	31						232	233						18	
209		46											35		224
			61			199			202			52			
			180	76							69	189			
	162				91					86				175	
					150	106			103	155					
		131					121	120					142		
		115					137	136					126		
					102	154			151	107					
	82				171					166				95	
			68	188							181	77			
			205			55			58			196			
33		222											211		48
	239						24	25						226	
256				5							12				241

Magic Square
Number Fill-ins [B] [3]
Mathematical Formula 16 X 16 + 1 X 8 = 2056

SHELSEAS MAGIC SQUARE
Magical Stars 2056

Positive [B] Start with the top row, right to left in each row, going down. Begin with the number 1, count each square, only fill-in same shaded squares. Diagonals are shown for the same sum of 2056.

16															1
	31													18	
		46											35		
			61									52			
				76							69				
					91					86					
						106			103						
							121	120							
							137	136							
						154			151						
					171					166					
				188							181				
			205									196			
		222											211		
	239													226	
256															241

Negative [B] Start with the bottom row, left to right in each row, going up. Begin with the number 3, count each square, only fill-in same shaded squares.

		243												254	
				231			234								
					216	217									
			197							204					
			180								189				
161															176
				150						155					
	130													143	
	114													127	
				102						107					
81															96
			68								77				
				53							60				
						40	41								
					23			26							
		3												14	

Magic Square - Number Fill-ins [B] [4] - Mathematical Formula 16 X 16 + 1 X 8 = 2056

SHELSEAS MAGIC SQUARE
Magical Stars 2056

Positive [B] Start with the top row, right to left in each row, going down. Begin with the number 1, count each square, only fill-in same shaded squares. Diagonals are shown for the same sum of 2056.

Negative [B] Start with the bottom row, left to right in each row, going up. Begin with the number 3, count each square, only fill-in same shaded squares.

Objective: Combine both Positive & Negative number fill-ins [B] [4]. Add each square of numbers in any row, column, or diagonal to proof the same sum of 2056 for a solved Magic Square. Helpful hints below.

16		243											254		1
	31				231			234					18		
		46				216	217					35			
			61	197							204	52			
			180	76							69	189			
161					91				86						176
					150	106			103	155					
	130					121	120						143		
	114					137	136						127		
					102	154			151	107					
81					171				166						96
			68	188							181	77			
			205	53							60	196			
		222				40	41						211		
	239				23			26					226		
256		3											14		241

Magic Square
Number Fill-ins [B] [4]
Mathematical Formula 16 X 16 + 1 X 8 = 2056

SHELSEAS MAGIC SQUARE
Magical Stars 2056

Positive [B] Start with the top row, right to left in each row, going down. Begin with the number 1, count each square, only fill-in same shaded squares. Diagonals are shown for the same sum of 2056.

16															1
	31													18	
		46											35		
			61									52			
				76							69				
					91					86					
						106			103						
							121	120							
							137	136							
						154			151						
					171					166					
				188							181				
			205									196			
		222											211		
	239													226	
256															241

Negative [B] Start with the bottom row, left to right in each row, going up. Begin with the number 4, count each square, only fill-in same shaded squares.

			244									253			
				229							236				
					214					219					
						200	201								
177															192
	162													175	
		147											158		
						135			138						
						119			122						
		99											110		
	82													95	
65															80
						56	57								
					38					43					
				21							28				
			4									13			

Magic Square - Number Fill-ins [B] [5] - Mathematical Formula 16 X 16 + 1 X 8 = 2056

SHELSEAS MAGIC SQUARE
Magical Stars 2056

Positive [B] Start with the top row, right to left in each row, going down. Begin with the number 1, count each square, only fill-in same shaded squares. Diagonals are shown for the same sum of 2056.

Negative [B] Start with the bottom row, left to right in each row, going up. Begin with the number 4, count each square, only fill-in same shaded squares.

Objective: Combine both Positive & Negative number fill-ins [B] [5]. Add each square of numbers in any row, column, or diagonal to proof the same sum of 2056 for a solved Magic Square. Helpful hints below.

16			244									253			1
	31			229							236			18	
		46			214					219			35		
			61			200	201				52				
177			76								69				192
	162				91					86				175	
		147			106			103				158			
					135	121	120	138							
					119	137	136	122							
		99			154			151				110			
	82				171					166				95	
65			188								181				80
		205				56	57				196				
		222		38					43			211			
	239			21						28			226		
256			4								13				241

Magic Square
Number Fill-ins [B] [5]
Mathematical Formula 16 X 16 + 1 X 8 = 2056

Solved

[B]

Magic Square
Number Fill-ins [B]
Mathematical Formula 16 X 16 + 1 X 8 = 2056

SHELSEAS MAGIC SQUARE
Magical Stars 2056

Positive [B] Diagonals are shown for the same sum of 2056

16	15	14	13									4	3	2	1
32	31		29		26		23			20		18	17		
48		46	45		43					38		36	35		33
64	63	62	61									52	51	50	49
				76	75	74	73	72	71	70	69				
		94		92	91		89	88		86	85		83		
	111			108		106	105	104	103		101			98	
				124	123	122	121	120	119	118	117				
				140	139	138	137	136	135	134	133				
	159			156		154	153	152	151		149			146	
		174		172	171		169	168		166	165		163		
				188	187	186	185	184	183	182	181				
208	207	206	205									196	195	194	193
224		222	221		219				214			212	211		209
240	239		237		234			231				228		226	225
256	255	254	253									244	243	242	241

Negative [B]

				245	246	247	248	249	250	251	252				
		227		229	230		232	233		235	236		238		
	210			213		215	216	217	218		220			223	
				197	198	199	200	201	202	203	204				
177	178	179	180									189	190	191	192
161	162		164			167			170			173		175	176
145		147	148		150					155		157	158		160
129	130	131	132									141	142	143	144
113	114	115	116									125	126	127	128
97		99	100		102					107		109	110		112
81	82		84			87			90			93		95	96
65	66	67	68									77	78	79	80
				53	54	55	56	57	58	59	60				
	34			37		39	40	41	42		44			47	
		19		21	22		24	25		27	28		30		
				5	6	7	8	9	10	11	12				

Magic Square
Number Fill-ins [B] [1]
Mathematical Formula 16 X 16 + 1 X 8 = 2056

SHELSEAS MAGIC SQUARE
Magical Stars 2056

Combination of both Positive & Negative number fill-ins [B] [1] is shown for a solved Magic Square. Objective: Add each square of numbers in any row, column, or diagonal to proof the same sum of 2056.

16	15	14	13	245	246	247	248	249	250	251	252	4	3	2	1
32	31	227	29	229	230	26	232	233	23	235	236	20	238	18	17
48	210	46	45	213	43	215	216	217	218	38	220	36	35	223	33
64	63	62	61	197	198	199	200	201	202	203	204	52	51	50	49
177	178	179	180	76	75	74	73	72	71	70	69	189	190	191	192
161	162	94	164	92	91	167	89	88	170	86	85	173	83	175	176
145	111	147	148	108	150	106	105	104	103	155	101	157	158	98	160
129	130	131	132	124	123	122	121	120	119	118	117	141	142	143	144
113	114	115	116	140	139	138	137	136	135	134	133	125	126	127	128
97	159	99	100	156	102	154	153	152	151	107	149	109	110	146	112
81	82	174	84	172	171	87	169	168	90	166	165	93	163	95	96
65	66	67	68	188	187	186	185	184	183	182	181	77	78	79	80
208	207	206	205	53	54	55	56	57	58	59	60	196	195	194	193
224	34	222	221	37	219	39	40	41	42	214	44	212	211	47	209
240	239	19	237	21	22	234	24	25	231	27	28	228	30	226	225
256	255	254	253	5	6	7	8	9	10	11	12	244	243	242	241

2056

Magic Square
Number Fill-ins [B] [1]
Mathematical Formula 16 X 16 + 1 X 8 = 2056

SHELSEAS MAGIC SQUARE
Magical Stars 2056

Positive [B] Diagonals are the shown for the same sum of 2056

16	15				11		9	8		6				2	1
32	31		29			26			23			20		18	17
		46		44	43	42			39	38	37		35		
	63		61	60			57	56			53	52		50	
		78	77	76			73	72			69	68	67		
96		94			91	90			87	86			83		81
	111	110			107	106			103	102			99	98	
128			125	124			121	120			117	116			113
144			141	140			137	136			133	132			129
	159	158			155	154			151	150			147	146	
176		174			171	170			167	166			163		161
		190	189	188			185	184			181	180	179		
	207		205	204			201	200			197	196		194	
		222		220	219	218			215	214	213		211		
240	239		237			234			231			228		226	225
256	255				251		249	248		246				242	241

Negative [B]

		243	244	245		247			250		252	253	254		
		227		229	230		232	233		235	236		238		
209	210		212				216	217				221		223	224
193		195			198	199			202	203			206		208
177	178				182	183			186	187				191	192
	162		164	165			168	169			172	173		175	
145			148	149			152	153			156	157			160
	130	131			134	135			138	139			142	143	
	114	115			118	119			122	123			126	127	
97			100	101			104	105			108	109			112
	82		84	85			88	89			92	93		95	
65	66				70	71			74	75				79	80
49		51			54	55			58	59			62		64
33	34		36				40	41				45		47	48
		19		21	22		24	25		27	28		30		
		3	4	5		7			10		12	13	14		

Magic Square
Number Fill-ins [B] [2]
Mathematical Formula 16 X 16 + 1 X 8 = 2056

SHELSEAS MAGIC SQUARE
Magical Stars 2056

Combination of both Positive & Negative number fill-ins [B] [2] is shown for a solved Magic Square. Objective: Add each square of numbers in any row, column, or diagonal to proof the same sum of 2056.

16	15	243	244	245	11	247	9	8	250	6	252	253	254	2	1
32	31	227	29	229	230	26	232	233	23	235	236	20	238	18	17
209	210	46	212	44	43	42	216	217	39	38	37	221	35	223	224
193	63	195	61	60	198	199	57	56	202	203	53	52	206	50	208
177	178	78	77	76	182	183	73	72	186	187	69	68	67	191	192
96	162	94	164	165	91	90	168	169	87	86	172	173	83	175	81
145	111	110	148	149	107	106	152	153	103	102	156	157	99	98	160
128	130	131	125	124	134	135	121	120	138	139	117	116	142	143	113
144	114	115	141	140	118	119	137	136	122	123	133	132	126	127	129
97	159	158	100	101	155	154	104	105	151	150	108	109	147	146	112
176	82	174	84	85	171	170	88	89	167	166	92	93	163	95	161
65	66	190	189	188	70	71	185	184	74	75	181	180	179	79	80
49	207	51	205	204	54	55	201	200	58	59	197	196	62	194	64
33	34	222	36	220	219	218	40	41	215	214	213	45	211	47	48
240	239	19	237	21	22	234	24	25	231	27	28	228	30	226	225
256	255	3	4	5	251	7	249	248	10	246	12	13	14	242	241

2056

Magic Square
Number Fill-ins [B] [2]
Mathematical Formula 16 X 16 + 1 X 8 = 2056

SHELSEAS MAGIC SQUARE
Magical Stars 2056

Positive [B] Diagonals are shown for the same sum of 2056

16	15	14	13									4	3	2	1
32	31	30	29									20	19	18	17
		46	45	44	43					38	37	36	35		
		62	61	60	59					54	53	52	51		
			76	75	74	73	72	71	70	69					
			92	91	90	89	88	87	86	85					
112	111				106	105	104	103						98	97
128	127				122	121	120	119						114	113
144	143				138	137	136	135						130	129
160	159				154	153	152	151						146	145
				172	171	170	169	168	167	166	165				
				188	187	186	185	184	183	182	181				
		206	205	204	203					198	197	196	195		
		222	221	220	219					214	213	212	211		
240	239	238	237									228	227	226	225
256	255	254	253									244	243	242	241

Negative [B]

				245	246	247	248	249	250	251	252				
				229	230	231	232	233	234	235	236				
209	210				215	216	217	218						223	224
193	194				199	200	201	202						207	208
177	178	179	180									189	190	191	192
161	162	163	164									173	174	175	176
		147	148	149	150					155	156	157	158		
		131	132	133	134					139	140	141	142		
		115	116	117	118					123	124	125	126		
		99	100	101	102					107	108	109	110		
81	82	83	84									93	94	95	96
65	66	67	68									77	78	79	80
49	50				55	56	57	58						63	64
33	34				39	40	41	42						47	48
				21	22	23	24	25	26	27	28				
				5	6	7	8	9	10	11	12				

Magic Square
Number Fill-ins [B] [3]
Mathematical Formula 16 X 16 + 1 X 8 = 2056

SHELSEAS MAGIC SQUARE
Magical Stars 2056

Combination of both Positive & Negative number fill-ins [B] [3] is shown for a solved Magic Square. Objective: Add each square of numbers in any row, column, or diagonal to proof the same sum of 2056.

16	15	14	13	245	246	247	248	249	250	251	252	4	3	2	1
32	31	30	29	229	230	231	232	233	234	235	236	20	19	18	17
209	210	46	45	44	43	215	216	217	218	38	37	36	35	223	224
193	194	62	61	60	59	199	200	201	202	54	53	52	51	207	208
177	178	179	180	76	75	74	73	72	71	70	69	189	190	191	192
161	162	163	164	92	91	90	89	88	87	86	85	173	174	175	176
112	111	147	148	149	150	106	105	104	103	155	156	157	158	98	97
128	127	131	132	133	134	122	121	120	119	139	140	141	142	114	113
144	143	115	116	117	118	138	137	136	135	123	124	125	126	130	129
160	159	99	100	101	102	154	153	152	151	107	108	109	110	146	145
81	82	83	84	172	171	170	169	168	167	166	165	93	94	95	96
65	66	67	68	188	187	186	185	184	183	182	181	77	78	79	80
49	50	206	205	204	203	55	56	57	58	198	197	196	195	63	64
33	34	222	221	220	219	39	40	41	42	214	213	212	211	47	48
240	239	238	237	21	22	23	24	25	26	27	28	228	227	226	225
256	255	254	253	5	6	7	8	9	10	11	12	244	243	242	241

2056

Magic Square
Number Fill-ins [B] [3]
Mathematical Formula 16 X 16 + 1 X 8 = 2056

SHELSEAS MAGIC SQUARE
Magical Stars 2056

Positive [B] Diagonals are shown for the same sum of 2056

16	15		13				9	8				4		2	1
32	31		29		27					22		20		18	17
		46	45		43	42			39	38		36	35		
64	63	62	61									52	51	50	49
				76	75	74	73	72	71	70	69				
	95	94		92	91					86	85		83	82	
		110		108		106	105	104	103		101		99		
128				124		122	121	120	119		117				113
144				140		138	137	136	135		133				129
		158		156		154	153	152	151		149		147		
	175	174		172	171					166	165		163	162	
				188	187	186	185	184	183	182	181				
208	207	206	205									196	195	194	193
		222	221		219	218			215	214		212	211		
240	239		237		235					230		228		226	225
256	255		253				249	248				244		242	241

Negative [B]

		243		245	246	247			250	251	252		254		
		227		229		231	232	233	234		236		238		
209	210			213			216	217			220			223	224
				197	198	199	200	201	202	203	204				
177	178	179	180									189	190	191	192
161			164			167	168	169	170			173			176
145	146		148		150					155		157		159	160
	130	131	132		134					139		141	142	143	
	114	115	116		118					123		125	126	127	
97	98		100		102					107		109		111	112
81			84			87	88	89	90			93			96
65	66	67	68									77	78	79	80
				53	54	55	56	57	58	59	60				
33	34			37			40	41			44			47	48
		19		21		23	24	25	26		28		30		
		3		5	6	7			10	11	12		14		

Magic Square
Number Fill-ins [B] [4]
Mathematical Formula 16 X 16 + 1 X 8 = 2056

SHELSEAS MAGIC SQUARE
Magical Stars 2056

Combination of both Positive & Negative number fill-ins [B] [4] is shown for a solved Magic Square. Objective: Add each square of numbers in any row, column, or diagonal to proof the same sum of 2056.

16	15	243	13	245	246	247	9	8	250	251	252	4	254	2	1
32	31	227	29	229	27	231	232	233	234	22	236	20	238	18	17
209	210	46	45	213	43	42	216	217	39	38	220	36	35	223	224
64	63	62	61	197	198	199	200	201	202	203	204	52	51	50	49
177	178	179	180	76	75	74	73	72	71	70	69	189	190	191	192
161	95	94	164	92	91	167	168	169	170	86	85	173	83	82	176
145	146	110	148	108	150	106	105	104	103	155	101	157	99	159	160
128	130	131	132	124	134	122	121	120	119	139	117	141	142	143	113
144	114	115	116	140	118	138	137	136	135	123	133	125	126	127	129
97	98	158	100	156	102	154	153	152	151	107	149	109	147	111	112
81	175	174	84	172	171	87	88	89	90	166	165	93	163	162	96
65	66	67	68	188	187	186	185	184	183	182	181	77	78	79	80
208	207	206	205	53	54	55	56	57	58	59	60	196	195	194	193
33	34	222	221	37	219	218	40	41	215	214	44	212	211	47	48
240	239	19	237	21	235	23	24	25	26	230	28	228	30	226	225
256	255	3	253	5	6	7	249	248	10	11	12	244	14	242	241

2056

Magic Square
Number Fill-ins [B] [4]
Mathematical Formula 16 X 16 + 1 X 8 = 2056

SHELSEAS MAGIC SQUARE
Magical Stars 2056

Positive [B] Diagonals are shown for the same sum of 2056

16	15	14			10			7					3	2	1
32	31	30				26			23				19	18	17
48	47	46	45									36	35	34	33
	63	62	61	60							53	52	51	50	
			77	76	75		73	72		70	69	68			
				92	91	90	89	88	87	86	85				
112					107	106	105	104	103	102					97
			125	124	123		121	120		118	117	116			
			141	140	139		137	136		134	133	132			
160					155	154	153	152	151	150					145
				172	171	170	169	168	167	166	165				
			189	188	187		185	184		182	181	180			
	207	206	205	204							197	196	195	194	
224	223	222	221									212	211	210	209
240	239	238				234			231				227	226	225
256	255	254				250			247				243	242	241

Negative [B]

			244	245	246		248	249		251	252	253			
			228	229	230		232	233		235	236	237			
			213	214	215	216	217	218	219	220					
193					198	199	200	201	202	203					208
177	178	179			183			186				190	191	192	
161	162	163	164								173	174	175	176	
	146	147	148	149							156	157	158	159	
129	130	131				135			138			142	143	144	
113	114	115				119			122			126	127	128	
	98	99	100	101						108	109	110	111		
81	82	83	84									93	94	95	96
65	66	67				71			74				78	79	80
49					54	55	56	57	58	59					64
			37	38	39	40	41	42	43	44					
			20	21	22		24	25		27	28	29			
			4	5	6		8	9		11	12	13			

Magic Square
Number Fill-ins [B] [5]
Mathematical Formula 16 X 16 + 1 X 8 = 2056

SHELSEAS MAGIC SQUARE
Magical Stars 2056

Combination of both Positive & Negative number fill-ins [B] [5] is shown for a solved Magic Square. Objective: Add each square of numbers in any row, column, or diagonal to proof the same sum of 2056.

16	15	14	244	245	246	10	248	249	7	251	252	253	3	2	1
32	31	30	228	229	230	26	232	233	23	235	236	237	19	18	17
48	47	46	45	213	214	215	216	217	218	219	220	36	35	34	33
193	63	62	61	60	198	199	200	201	202	203	53	52	51	50	208
177	178	179	77	76	75	183	73	72	186	70	69	68	190	191	192
161	162	163	164	92	91	90	89	88	87	86	85	173	174	175	176
112	146	147	148	149	107	106	105	104	103	102	156	157	158	159	97
129	130	131	125	124	123	135	121	120	138	118	117	116	142	143	144
113	114	115	141	140	139	119	137	136	122	134	133	132	126	127	128
160	98	99	100	101	155	154	153	152	151	150	108	109	110	111	145
81	82	83	84	172	171	170	169	168	167	166	165	93	94	95	96
65	66	67	189	188	187	71	185	184	74	182	181	180	78	79	80
49	207	206	205	204	54	55	56	57	58	59	197	196	195	194	64
224	223	222	221	37	38	39	40	41	42	43	44	212	211	210	209
240	239	238	20	21	22	234	24	25	231	27	28	29	227	226	225
256	255	254	4	5	6	250	8	9	247	11	12	13	243	242	241

2056

Magic Square
Number Fill-ins [B] [5]
Mathematical Formula 16 X 16 + 1 X 8 = 2056

The Number Fill-ins [C] Will Be Your Guideline
For A Solved Double Even – Magic Square.

Number
Fill-ins

[C]

Positive [C] Start with the bottom row, right to left in each row, going up. Begin with the number shown hint, count each square, only fill-in shaded squares.

Negative [C] Start with the top row, left to right in each row, going down. Begin with the number shown hint, count each square, only fill-in shaded squares.

Objective: Combine both Positive & Negative number fill-ins [C] in each Magic Square. Add each square of numbers in any row, column, or diagonal to proof the same sum of 2056 for a solved Magic Square.

SHELSEAS MAGIC SQUARE
Magical Stars 2056

Positive [C] Start with the bottom row, right to left in each row, going up. Begin with the number 1, count each square, only fill-in shaded squares.

16			13
	11	10	
	7	6	
4			1

Negative [C] Start with the top row, left to right in each row, going down. Begin with the number 2, count each square, only fill-in shaded squares.

	2	3	
5			8
9			12
	14	15	

Combination of both Positive & Negative number fill-ins [C] is shown for a solved Magic Square. Objective: Add each square of numbers in any row, column or diagonal to proof the same sum of 34.

16	2	3	13
5	11	10	8
9	7	6	12
4	14	15	1

Magic Square
Number Fill-ins [C]
Mathematical Formula 4 X 4 + 1 X 2 = 34

SHELSEAS MAGIC SQUARE
Magical Stars 2056

Positive [C] Start with the bottom row, right to left in each row, going up. Begin with the number 1, count each square, only fill-in shaded squares. Diagonals are shown for the same sum of 2056. Helpful hints below.

256															241
	239													226	
		222											211		
			205									196			
				188							181				
					171					166					
						154			151						
							137	136							
							121	120							
						106			103						
					91					86					
				76							69				
			61									52			
		46											35		
	31													18	
16															1

Negative [C] Start with the top row, left to right in each row, going down. Begin with the number 5, count each square, only fill-in shaded squares. Helpful hints below.

				5							12				
					23			26							
			36									45			
		51											62		
65															80
						88	89								
	98													111	
					118					123					
					134					139					
	146													159	
						168	169								
177															192
		195											206		
			212									221			
					231			234							
				245							252				

Magic Square - Number Fill-ins [C] [1] - Mathematical Formula 16 X 16 + 1 X 8 = 2056

SHELSEAS MAGIC SQUARE
Magical Stars 2056

Positive [C] Start with the bottom row, right to left in each row, going up. Begin with the number 1, count each square, only fill-in same shaded squares. Diagonals are shown for the same sum of 2056.

Negative [C] Start with the top row, left to right in each row, going down. Begin with the number 5, count each square, only fill-in same shaded squares.

Objective: Combine both Positive & Negative number fill-ins [C] [1]. Add each square of numbers in any row, column, or diagonal to proof the same sum of 2056 for a solved Magic Square. Helpful hints below.

256				5							12				241
	239				23			26						226	
		222	36									45	211		
		51	205									196	62		
65				188							181				80
					171		88	89		166					
	98					154			151					111	
					118		137	136		123					
					134		121	120		139					
	146					106			103					159	
					91		168	169		86					
177				76							69				192
		195	61									52	206		
		46	212									221	35		
	31				231			234						18	
16				245							252				1

Magic Square
Number Fill-ins [C] [1]
Mathematical Formula 16 X 16 + 1 X 8 = 2056

SHELSEAS MAGIC SQUARE
Magical Stars 2056

Positive [C] Start with the bottom row, right to left in each row, going up. Begin with the number 1, count each square, only fill-in shaded squares. Diagonals are shown for the same sum of 2056. Helpful hints below.

256															241
	239													226	
		222											211		
			205									196			
				188							181				
					171					166					
						154			151						
							137	136							
							121	120							
						106			103						
					91					86					
				76							69				
			61									52			
		46											35		
	31													18	
16															1

Negative [C] Start with the top row, left to right in each row, going down. Begin with the number 5, count each square, only fill-in shaded squares. Helpful hints below.

				5							12				
						23			26						
							40	41							
					54						59				
65															80
			84									93			
	98													111	
		115											126		
		131											142		
	146													159	
			164									173			
177															192
					198					203					
							216	217							
						231			234						
				245							252				

Magic Square - Number Fill-ins [C] [2] - Mathematical Formula 16 X 16 + 1 X 8 = 2056

SHELSEAS MAGIC SQUARE
Magical Stars 2056

Positive [C] Start with the bottom row, right to left in each row, going up. Begin with the number 1, count each square, only fill-in same shaded squares. Diagonals are shown for the same sum of 2056.

Negative [C] Start with the top row, left to right in each row, going down. Begin with the number 5, count each square, only fill-in same shaded squares.

Objective: Combine both Positive & Negative number fill-ins [C] [2]. Add each square of numbers in any row, column, or diagonal to proof the same sum of 2056 for a solved Magic Square. Helpful hints below.

256				5							12				241
	239					23			26					226	
		222					40	41					211		
			205		54					59		196			
65				188							181				80
			84		171					166		93			
	98					154			151					111	
		115					137	136					126		
		131					121	120					142		
	146					106			103					159	
			164		91					86		173			
177				76							69				192
			61		198					203		52			
		46					216	217					35		
	31					231			234					18	
16				245							252				1

Magic Square
Number Fill-ins [C] [2]
Mathematical Formula 16 X 16 + 1 X 8 = 2056

SHELSEAS MAGIC SQUARE
Magical Stars 2056

Positive [C] Start with the bottom row, right to left in each row, going up. Begin with the number 1, count each square, only fill-in shaded squares. Diagonals are shown for the same sum of 2056. Helpful hints below.

256															241
	239													226	
		222											211		
			205									196			
				188							181				
					171					166					
						154			151						
							137	136							
							121	120							
						106			103						
					91					86					
				76							69				
			61									52			
		46											35		
	31													18	
16															1

Negative [C] Start with the top row, left to right in each row, going down. Begin with the number 4, count each square, only fill-in shaded squares. Helpful hints below.

			4									13			
				22							27				
					39			42							
49															64
						72	73								
	82													95	
		99											110		
			117									124			
			133									140			
		147											158		
	162													175	
						184	185								
193															208
					215			218							
				230							235				
			244									253			

Magic Square - Number Fill-ins [C] [3] - Mathematical Formula 16 X 16 + 1 X 8 = 2056

SHELSEAS MAGIC SQUARE
Magical Stars 2056

Positive [C] Start with the bottom row, right to left in each row, going up. Begin with the number 1, count each square, only fill-in same shaded squares. Diagonals are shown for the same sum of 2056.

Negative [C] Start with the top row, left to right in each row, going down. Begin with the number 4, count each square, only fill-in same shaded squares.

Objective: Combine both Positive & Negative number fill-ins [C] [3]. Add each square of numbers in any row, column, or diagonal to proof the same sum of 2056 for a solved Magic Square. Helpful hints below.

256			4									13			241
	239				22					27				226	
		222				39			42				211		
49			205									196			64
				188			72	73			181				
	82				171					166				95	
		99				154			151				110		
				117			137	136			124				
				133			121	120			140				
		147				106			103				158		
	162				91					86				175	
				76			184	185			69				
193			61									52			208
		46				215			218				35		
	31				230					235				18	
16			244									253			1

Magic Square
Number Fill-ins [C] [3]
Mathematical Formula 16 X 16 + 1 X 8 = 2056

SHELSEAS MAGIC SQUARE
Magical Stars 2056

Positive [C] Start with the bottom row, right to left in each row, going up. Begin with the number 1, count each square, only fill-in shaded squares. Diagonals are shown for the same sum of 2056. Helpful hints below.

256															241
	239													226	
		222											211		
			205									196			
				188							181				
					171					166					
						154			151						
							137	136							
							121	120							
						106			103						
					91					86					
				76							69				
			61									52			
		46											35		
	31													18	
16															1

Negative [C] Start with the top row, left to right in each row, going down. Begin with the number 4, count each square, only fill-in same shaded squares.

			4								13				
					23			26							
						40	41								
				54						59					
		67										78			
	82												95		
97															112
			117							124					
			133							140					
145															160
	162												175		
		179										190			
				198					203						
						216	217								
					231			234							
			244								253				

Magic Square - Number Fill-ins [C] [4] - Mathematical Formula 16 X 16 + 1 X 8 = 2056

SHELSEAS MAGIC SQUARE
Magical Stars 2056

Positive [C] Start with the bottom row, right to left in each row, going up. Begin with the number 1, count each square, only fill-in same shaded squares. Diagonals are shown for the same sum of 2056.

Negative [C] Start with the top row, left to right in each row, going down. Begin with the number 4, count each square, only fill-in same shaded squares.

Objective: Combine both Positive & Negative number fill-ins [C] [4]. Add each square of numbers in any row, column, or diagonal to proof the same sum of 2056 for a solved Magic Square. Helpful hints below.

256			4										13		241
	239				23			26						226	
		222				40	41						211		
			205		54					59		196			
		67		188							181		78		
	82					171				166				95	
97						154		151							112
				117		137	136				124				
				133		121	120				140				
145						106		103							160
	162				91					86				175	
		179		76							69		190		
			61		198					203		52			
		46				216	217						35		
	31					231		234						18	
16			244									253			1

Magic Square
Number Fill-ins [C] [4]
Mathematical Formula 16 X 16 + 1 X 8 = 2056

SHELSEAS MAGIC SQUARE
Magical Stars 2056

Positive [C] Start with the bottom row, right to left in each row, going up. Begin with the number 1, count each square, only fill-in shaded squares. Diagonals are shown for the same sum of 2056. Helpful hints below.

256															241
	239													226	
		222											211		
			205									196			
				188							181				
					171					166					
						154			151						
							137	136							
							121	120							
						106			103						
					91					86					
				76							69				
			61									52			
		46											35		
	31													18	
16															1

Negative [C] Start with the top row, left to right in each row, going down. Begin with the number 4, count each square, only fill-in same shaded squares.

			4									13			
		19											30		
	34													47	
49															64
						72	73								
					87			90							
				102					107						
			117							124					
			133							140					
				150					155						
					167			170							
						184	185								
193															208
	210													223	
		227											238		
			244									253			

Magic Square - Number Fill-ins [C] [5] - Mathematical Formula 16 X 16 + 1 X 8 = 2056

SHELSEAS MAGIC SQUARE
Magical Stars 2056

Positive [C] Start with the bottom row, right to left in each row, going up. Begin with the number 1, count each square, only fill-in same shaded squares. Diagonals are shown for the same sum of 2056.

Negative [C] Start with the top row, left to right in each row, going down. Begin with the number 4, count each square, only fill-in same shaded squares.

Objective: Combine both Positive & Negative number fill-ins [C] [5]. Add each square of numbers in any row, column, or diagonal to proof the same sum of 2056 for a solved Magic Square. Helpful hints below.

256			4									13			241
	239	19											30	226	
	34	222											211	47	
49			205									196			64
			188			72	73				181				
				171	87			90	166						
				102	154			151	107						
			117			137	136				124				
			133			121	120				140				
				150	106			103	155						
				91	167			170	86						
			76			184	185				69				
193			61									52			208
	210	46											35	223	
	31	227											238	18	
16			244									253			1

Magic Square
Number Fill-ins [C] [5]
Mathematical Formula 16 X 16 + 1 X 8 = 2056

Solved

[C]

SHELSEAS MAGIC SQUARE
Magical Stars 2056

Positive [C] Diagonals are shown for the same sum of 2056

256	255	254	253									244	243	242	241
240	239			236	235					230	229			226	225
224		222			219	218			215	214			211		209
208			205			202	201	200	199			196			193
	191			188	187		185	184		182	181			178	
	175	174		172	171					166	165		163	162	
		158	157			154	153	152	151			148	147		
			141	140		138	137	136	135		133	132			
			125	124		122	121	120	119		117	116			
		110	109			106	105	104	103			100	99		
	95	94		92	91					86	85		83	82	
	79			76	75		73	72		70	69			66	
64			61			58	57	56	55			52			49
48		46			43	42			39	38			35		33
32	31			28	27					22	21			18	17
16	15	14	13									4	3	2	1

Negative [C]

				5	6	7	8	9	10	11	12				
		19	20			23	24	25	26			29	30		
	34		36	37			40	41			44	45		47	
	50	51		53	54					59	60		62	63	
65		67	68			71			74			77	78		80
81			84			87	88	89	90			93			96
97	98			101	102					107	108			111	112
113	114	115			118					123			126	127	128
129	130	131			134					139			142	143	144
145	146			149	150					155	156			159	160
161			164			167	168	169	170			173			176
177		179	180			183			186			189	190		192
	194	195		197	198					203	204		206	207	
	210		212	213			216	217			220	221		223	
		227	228			231	232	233	234			237	238		
				245	246	247	248	249	250	251	252				

Magic Square
Number Fill-ins [C] [1]
Mathematical Formula 16 X 16 + 1 X 8 = 2056

SHELSEAS MAGIC SQUARE
Magical Stars 2056

Combination of both Positive & Negative number fill-ins [C] [1] is shown for a solved Magic Square. Objective: Add each square of numbers in any row, column, or diagonal to proof the same sum of 2056.

256	255	254	253	5	6	7	8	9	10	11	12	244	243	242	241
240	239	19	20	236	235	23	24	25	26	230	229	29	30	226	225
224	34	222	36	37	219	218	40	41	215	214	44	45	211	47	209
208	50	51	205	53	54	202	201	200	199	59	60	196	62	63	193
65	191	67	68	188	187	71	185	184	74	182	181	77	78	178	80
81	175	174	84	172	171	87	88	89	90	166	165	93	163	162	96
97	98	158	157	101	102	154	153	152	151	107	108	148	147	111	112
113	114	115	141	140	118	138	137	136	135	123	133	132	126	127	128
129	130	131	125	124	134	122	121	120	119	139	117	116	142	143	144
145	146	110	109	149	150	106	105	104	103	155	156	100	99	159	160
161	95	94	164	92	91	167	168	169	170	86	85	173	83	82	176
177	79	179	180	76	75	183	73	72	186	70	69	189	190	66	192
64	194	195	61	197	198	58	57	56	55	203	204	52	206	207	49
48	210	46	212	213	43	42	216	217	39	38	220	221	35	223	33
32	31	227	228	28	27	231	232	233	234	22	21	237	238	18	17
16	15	14	13	245	246	247	248	249	250	251	252	4	3	2	1

2056

Magic Square
Number Fill-ins [C] [1]
Mathematical Formula 16 X 16 + 1 X 8 = 2056

SHELSEAS MAGIC SQUARE
Magical Stars 2056

Positive [C] Diagonals are shown for the same sum of 2056

256	255	254	253									244	243	242	241
240	239	238					233	232					227	226	225
224	223	222	221									212	211	210	209
208		206	205	204							197	196	195		193
			189	188	187	186			183	182	181	180			
				172	171	170	169	168	167	166	165				
				156	155	154	153	152	151	150	149				
	143				139	138	137	136	135	134				130	
	127				123	122	121	120	119	118				114	
				108	107	106	105	104	103	102	101				
				92	91	90	89	88	87	86	85				
			77	76	75	74			71	70	69	68			
64		62	61	60							53	52	51		49
48	47	46	45									36	35	34	33
32	31	30					25	24					19	18	17
16	15	14	13									4	3	2	1

Negative [C]

				5	6	7	8	9	10	11	12				
			20	21	22	23			26	27	28	29			
				37	38	39	40	41	42	43	44				
	50				54	55	56	57	58	59				63	
65	66	67					72	73					78	79	80
81	82	83	84									93	94	95	96
97	98	99	100									109	110	111	112
113		115	116	117							124	125	126		128
129		131	132	133							140	141	142		144
145	146	147	148									157	158	159	160
161	162	163	164									173	174	175	176
177	178	179					184	185					190	191	192
	194				198	199	200	201	202	203				207	
				213	214	215	216	217	218	219	220				
			228	229	230	231			234	235	236	237			
				245	246	247	248	249	250	251	252				

Magic Square
Number Fill-ins [C] [2]
Mathematical Formula $16 \times 16 + 1 \times 8 = 2056$

SHELSEAS MAGIC SQUARE
Magical Stars 2056

Combination of both Positive & Negative number fill-ins [C] [2] is shown for a solved Magic Square. Objective: Add each square of numbers in any row, column, or diagonal to proof the same sum of 2056.

256	255	254	253	5	6	7	8	9	10	11	12	244	243	242	241
240	239	238	20	21	22	23	233	232	26	27	28	29	227	226	225
224	223	222	221	37	38	39	40	41	42	43	44	212	211	210	209
208	50	206	205	204	54	55	56	57	58	59	197	196	195	63	193
65	66	67	189	188	187	186	72	73	183	182	181	180	78	79	80
81	82	83	84	172	171	170	169	168	167	166	165	93	94	95	96
97	98	99	100	156	155	154	153	152	151	150	149	109	110	111	112
113	143	115	116	117	139	138	137	136	135	134	124	125	126	130	128
129	127	131	132	133	123	122	121	120	119	118	140	141	142	114	144
145	146	147	148	108	107	106	105	104	103	102	101	157	158	159	160
161	162	163	164	92	91	90	89	88	87	86	85	173	174	175	176
177	178	179	77	76	75	74	184	185	71	70	69	68	190	191	192
64	194	62	61	60	198	199	200	201	202	203	53	52	51	207	49
48	47	46	45	213	214	215	216	217	218	219	220	36	35	34	33
32	31	30	228	229	230	231	25	24	234	235	236	237	19	18	17
16	15	14	13	245	246	247	248	249	250	251	252	4	3	2	1

2056

Magic Square
Number Fill-ins [C] [2]
Mathematical Formula 16 X 16 + 1 X 8 = 2056

SHELSEAS MAGIC SQUARE
Magical Stars 2056

Positive [C] Diagonals are shown for the same sum of 2056

256	255	254					249	248					243	242	241
240	239	238	237									228	227	226	225
224	223	222	221									212	211	210	209
	207	206	205	204							197	196	195	194	
			189	188	187	186			183	182	181	180			
				172	171	170	169	168	167	166	165				
				156	155	154	153	152	151	150	149				
144					139	138	137	136	135	134					129
128					123	122	121	120	119	118					113
				108	107	106	105	104	103	102	101				
				92	91	90	89	88	87	86	85				
			77	76	75	74			71	70	69	68			
	63	62	61	60							53	52	51	50	
48	47	46	45									36	35	34	33
32	31	30	29									20	19	18	17
16	15	14					9	8					3	2	1

Negative [C]

			4	5	6	7			10	11	12	13			
				21	22	23	24	25	26	27	28				
				37	38	39	40	41	42	43	44				
49					54	55	56	57	58	59					64
65	66	67					72	73					78	79	80
81	82	83	84									93	94	95	96
97	98	99	100									109	110	111	112
	114	115	116	117							124	125	126	127	
	130	131	132	133							140	141	142	143	
145	146	147	148									157	158	159	160
161	162	163	164									173	174	175	176
177	178	179					184	185					190	191	192
193					198	199	200	201	202	203					208
				213	214	215	216	217	218	219	220				
				229	230	231	232	233	234	235	236				
			244	245	246	247			250	251	252	253			

Magic Square
Number Fill-ins [C] [3]
Mathematical Formula 16 X 16 + 1 X 8 = 2056

SHELSEAS MAGIC SQUARE
Magical Stars 2056

Combination of both Positive & Negative number fill-ins [C] [3] is shown for a solved Magic Square. Objective: Add each square of numbers in any row, column, or diagonal to proof the same sum of 2056.

256	255	254	4	5	6	7	249	248	10	11	12	13	243	242	241
240	239	238	237	21	22	23	24	25	26	27	28	228	227	226	225
224	223	222	221	37	38	39	40	41	42	43	44	212	211	210	209
49	207	206	205	204	54	55	56	57	58	59	197	196	195	194	64
65	66	67	189	188	187	186	72	73	183	182	181	180	78	79	80
81	82	83	84	172	171	170	169	168	167	166	165	93	94	95	96
97	98	99	100	156	155	154	153	152	151	150	149	109	110	111	112
144	114	115	116	117	139	138	137	136	135	134	124	125	126	127	129
128	130	131	132	133	123	122	121	120	119	118	140	141	142	143	113
145	146	147	148	108	107	106	105	104	103	102	101	157	158	159	160
161	162	163	164	92	91	90	89	88	87	86	85	173	174	175	176
177	178	179	77	76	75	74	184	185	71	70	69	68	190	191	192
193	63	62	61	60	198	199	200	201	202	203	53	52	51	50	208
48	47	46	45	213	214	215	216	217	218	219	220	36	35	34	33
32	31	30	29	229	230	231	232	233	234	235	236	20	19	18	17
16	15	14	244	245	246	247	9	8	250	251	252	253	3	2	1

2056

Magic Square
Number Fill-ins [C] [3]
Mathematical Formula 16 X 16 + 1 X 8 = 2056

SHELSEAS MAGIC SQUARE
Magical Stars 2056

Positive [C] Diagonals are shown for the same sum of 2056

256	255	254					249	248					243	242	241
240	239	238					233	232					227	226	225
224	223	222	221									212	211	210	209
		206	205	204		202			199		197	196	195		
			189	188	187	186			183	182	181	180			
				172	171	170	169	168	167	166	165				
			157	156	155	154			151	150	149	148			
144	143				139		137	136		134				130	129
128	127				123		121	120		118				114	113
			109	108	107	106			103	102	101	100			
				92	91	90	89	88	87	86	85				
			77	76	75	74			71	70	69	68			
		62	61	60		58			55		53	52	51		
48	47	46	45									36	35	34	33
32	31	30					25	24					19	18	17
16	15	14					9	8					3	2	1

Negative [C]

			4	5	6	7			10	11	12	13			
			20	21	22	23			26	27	28	29			
				37	38	39	40	41	42	43	44				
49	50				54		56	57		59				63	64
65	66	67					72	73					78	79	80
81	82	83	84									93	94	95	96
97	98	99					104	105					110	111	112
		115	116	117		119			122		124	125	126		
		131	132	133		135			138		140	141	142		
145	146	147					152	153					158	159	160
161	162	163	164									173	174	175	176
177	178	179					184	185					190	191	192
193	194				198		200	201		203				207	208
				213	214	215	216	217	218	219	220				
			228	229	230	231			234	235	236	237			
			244	245	246	247			250	251	252	253			

Magic Square
Number Fill-ins [C] [4]
Mathematical Formula 16 X 16 + 1 X 8 = 2056

SHELSEAS MAGIC SQUARE
Magical Stars 2056

Combination of both Positive & Negative number fill-ins [C] [4] is shown for a solved Magic Square. Objective: Add each square of numbers in any row, column, or diagonal to proof the same sum of 2056.

256	255	254	4	5	6	7	249	248	10	11	12	13	243	242	241
240	239	238	20	21	22	23	233	232	26	27	28	29	227	226	225
224	223	222	221	37	38	39	40	41	42	43	44	212	211	210	209
49	50	206	205	204	54	202	56	57	199	59	197	196	195	63	64
65	66	67	189	188	187	186	72	73	183	182	181	180	78	79	80
81	82	83	84	172	171	170	169	168	167	166	165	93	94	95	96
97	98	99	157	156	155	154	104	105	151	150	149	148	110	111	112
144	143	115	116	117	139	119	137	136	122	134	124	125	126	130	129
128	127	131	132	133	123	135	121	120	138	118	140	141	142	114	113
145	146	147	109	108	107	106	152	153	103	102	101	100	158	159	160
161	162	163	164	92	91	90	89	88	87	86	85	173	174	175	176
177	178	179	77	76	75	74	184	185	71	70	69	68	190	191	192
193	194	62	61	60	198	58	200	201	55	203	53	52	51	207	208
48	47	46	45	213	214	215	216	217	218	219	220	36	35	34	33
32	31	30	228	229	230	231	25	24	234	235	236	237	19	18	17
16	15	14	244	245	246	247	9	8	250	251	252	253	3	2	1

2056

Magic Square
Number Fill-ins [C] [4]
Mathematical Formula 16 X 16 + 1 X 8 = 2056

SHELSEAS MAGIC SQUARE
Magical Stars 2056

Positive [C] Diagonals are shown for the same sum of 2056

256	255	254				250			247				243	242	241
240	239				235	234			231	230				226	225
224		222	221	220							213	212	211		209
		206	205	204			201	200			197	196	195		
		190	189	188	187					182	181	180	179		
	175			172	171		169	168		166	165			162	
160	159					154	153	152	151					146	145
			141		139	138	137	136	135	134		132			
			125		123	122	121	120	119	118		116			
112	111					106	105	104	103					98	97
	95			92	91		89	88		86	85			82	
		78	77	76	75					70	69	68	67		
		62	61	60			57	56			53	52	51		
48		46	45	44							37	36	35		33
32	31				27	26			23	22				18	17
16	15	14				10			7				3	2	1

Negative [C]

			4	5	6		8	9		11	12	13			
		19	20	21			24	25			28	29	30		
	34				38	39	40	41	42	43				47	
49	50				54	55			58	59				63	64
65	66					71	72	73	74					79	80
81		83	84			87			90			93	94		96
		99	100	101	102					107	108	109	110		
113	114	115		117							124		126	127	128
129	130	131		133							140		142	143	144
		147	148	149	150					155	156	157	158		
161		163	164			167			170			173	174		176
177	178					183	184	185	186					191	192
193	194				198	199			202	203				207	208
	210				214	215	216	217	218	219				223	
		227	228	229			232	233			236	237	238		
			244	245	246		248	249		251	252	253			

Magic Square
Number Fill-ins [C] [5]
Mathematical Formula 16 X 16 + 1 X 8 = 2056

SHELSEAS MAGIC SQUARE
Magical Stars 2056

Combination of both Positive & Negative number fill-ins [C] [5] is shown for a solved
Magic Square. Objective: Add each square of numbers in any row, column, or diagonal
to proof the same sum of 2056.

256	255	254	4	5	6	250	8	9	247	11	12	13	243	242	241
240	239	19	20	21	235	234	24	25	231	230	28	29	30	226	225
224	34	222	221	220	38	39	40	41	42	43	213	212	211	47	209
49	50	206	205	204	54	55	201	200	58	59	197	196	195	63	64
65	66	190	189	188	187	71	72	73	74	182	181	180	179	79	80
81	175	83	84	172	171	87	169	168	90	166	165	93	94	162	96
160	159	99	100	101	102	154	153	152	151	107	108	109	110	146	145
113	114	115	141	117	139	138	137	136	135	134	124	132	126	127	128
129	130	131	125	133	123	122	121	120	119	118	140	116	142	143	144
112	111	147	148	149	150	106	105	104	103	155	156	157	158	98	97
161	95	163	164	92	91	167	89	88	170	86	85	173	174	82	176
177	178	78	77	76	75	183	184	185	186	70	69	68	67	191	192
193	194	62	61	60	198	199	57	56	202	203	53	52	51	207	208
48	210	46	45	44	214	215	216	217	218	219	37	36	35	223	33
32	31	227	228	229	27	26	232	233	23	22	236	237	238	18	17
16	15	14	244	245	246	10	248	249	7	251	252	253	3	2	1

2056

Magic Square
Number Fill-ins [C] [5]
Mathematical Formula 16 X 16 + 1 X 8 = 2056

The Number Fill-ins [D] Will Be Your Guideline
For A Solved Double Even – Magic Square.

Number
Fill-ins

[D]

Positive [D] Start with the bottom row, left to right in each row, going up. Begin with the number shown hint, count each square, only fill-in shaded squares.

Negative [D] Start with the top row, right to left in each row, going down. Begin with the number shown hint, count each square, only fill-in shaded squares.

Objective: Combine both Positive & Negative number fill-ins [D] in each Magic Square. Add each square of numbers in any row, column, or diagonal to proof the same sum of 2056 for a solved Magic Square.

SHELSEAS MAGIC SQUARE
Magical Stars 2056

Positive [D] Start with the bottom row, left to right in each row, going up. Begin with the number shown 1, count each square, only fill-in shaded squares.

13			16
	10	11	
	6	7	
1			4

Negative [D] Start with the top row, right to left in each row, going down. Begin with the number 2, count each square, only fill-in shaded squares.

	3	2	
8			5
12			9
	15	14	

Combination of both Positive & Negative number fill-ins [D] is shown for a solved Magic Square. Objective: Add each square of numbers in any row, column, or diagonal to proof the same sum of 34.

13	3	2	16
8	10	11	5
12	6	7	9
1	15	14	4

Magic Square
Number Fill-ins [D]
Mathematical Formula 4 X 4 + 1 X 2 = 34

SHELSEAS MAGIC SQUARE
Magical Stars 2056

Positive [D] Start with the bottom row, left to right in each row, going up. Begin with the number 1, count each square, only fill-in shaded squares. Diagonals are shown for the same sum of 2056. Helpful hints below.

241															256
	226													239	
		211											222		
			196									205			
				181							188				
					166					171					
						151			154						
							136	137							
							120	121							
						103			106						
					86					91					
				69							76				
			52									61			
		35											46		
	18													31	
1															16

Negative [D] Start with the top row, right to left in each row, going down. Begin with the number 3, count each square, only fill-in shaded squares. Helpful hints below.

		14											3		
			29									20			
48															33
	63													50	
						74			71						
							89	88							
				108							101				
					123					118					
					139					134					
				156							149				
							169	168							
						186			183						
	207													194	
224															209
			237									228			
		254											243		

Magic Square - Number Fill-ins [D] [1] - Mathematical Formula 16 X 16 + 1 X 8 = 2056

SHELSEAS MAGIC SQUARE
Magical Stars 2056

Positive [D] Start with the bottom row, left to right in each row, going up. Begin with the number 1, count each square, only fill-in same shaded squares. Diagonals are shown for the same sum of 2056.

Negative [D] Start with the top row, right to left in each row, going down. Begin with the number 3, count each square, only fill-in same shaded squares.

Objective: Combine both Positive & Negative number fill-ins [D] [1]. Add each square of numbers in any row, column, or diagonal to proof the same sum of 2056 for a solved Magic Square. Helpful hints below.

241		14											3		256
	226		29									20		239	
48		211											222		33
	63		196									205		50	
				181		74			71		188				
					166		89	88		171					
				108		151			154		101				
					123		136	137		118					
					139		120	121		134					
				156		103			106		149				
					86		169	168		91					
				69		186			183		76				
	207		52									61		194	
224		35											46		209
	18		237									228		31	
1		254											243		16

Magic Square
Number Fill-ins [D] [1]
Mathematical Formula 16 X 16 + 1 X 8 = 2056

SHELSEAS MAGIC SQUARE
Magical Stars 2056

Positive [D] Start with the bottom row, left to right in each row, going up. Begin with the number 1, count each square, only fill-in shaded squares. Diagonals are shown for the same sum of 2056. Helpful hints below.

241															256
	226													239	
		211											222		
			196									205			
				181							188				
					166					171					
						151			154						
							136	137							
							120	121							
						103			106						
					86					91					
				69							76				
			52									61			
		35											46		
	18													31	
1															16

NEGATIVE [D] Start with the top row, right to left in each row, going down. Begin with the number 5, count each square, only fill-in shaded squares. Helpful hints below.

				12							5				
						26			23						
	47													34	
							57	56							
80															65
		94											83		
					107					102					
			125									116			
			141									132			
					155					150					
		174											163		
192															177
						201	200								
	223													210	
						234			231						
				252							245				

Magic Square - Number Fill-ins [D] [2] - Mathematical Formula 16 X 16 + 1 X 8 = 2056

SHELSEAS MAGIC SQUARE
Magical Stars 2056

Positive [D] Start with the bottom row, left to right in each row, going up. Begin with the number 1, count each square, only fill-in same shaded squares. Diagonals are shown for the same sum of 2056.

Negative [D] Start with the top row, right to left in each row, going down. Begin with the number 5, count each square, only fill-in same shaded squares.

Objective: Combine both Positive & Negative number fill-ins [D] [2]. Add each square of numbers in any row, column, or diagonal to proof the same sum of 2056 for a solved Magic Square. Helpful hints below.

241				12							5				256
	226				26			23						239	
	47	211											222	34	
			196			57	56				205				
80				181							188				65
		94			166				171			83			
					107	151			154	102					
			125			136	137				116				
			141			120	121				132				
					155	103			106	150					
		174			86				91			163			
192				69							76				177
			52			201	200				61				
	223	35											46	210	
	18				234			231						31	
1				252							245				16

Magic Square
Number Fill-ins [D] [2]
Mathematical Formula 16 X 16 + 1 X 8 = 2056

SHELSEAS MAGIC SQUARE
Magical Stars 2056

Positive [D] Start with the bottom row, left to right in each row, going up. Begin with the number 1, count each square, only fill-in shaded squares. Diagonals are shown for the same sum of 2056. Helpful hints below.

241															256
	226													239	
		211											222		
			196									205			
				181							188				
					166					171					
						151			154						
							136	137							
							120	121							
						103			106						
					86					91					
				69							76				
			52									61			
		35											46		
	18													31	
1															16

Negative [D] Start with the top row, right to left in each row, going down. Begin with the number 3, count each square, only fill-in shaded squares. Helpful hints below.

		14												3	
					27					22					
48															33
						58			55						
	79													66	
						89	88								
				108							101				
			125									116			
			141									132			
				156							149				
						169	168								
	191													178	
						202			199						
224															209
					235					230					
		254											243		

Magic Square - Number Fill-ins [D] [3] - Mathematical Formula 16 X 16 + 1 X 8 = 2056

SHELSEAS MAGIC SQUARE
Magical Stars 2056

Positive [D] Start with the bottom row, left to right in each row, going up. Begin with the number 1, count each square, only fill-in same shaded squares. Diagonals are shown for the same sum of 2056.

Negative [D] Start with the top row, right to left in each row, going down. Begin with the number 3, count each square, only fill-in same shaded squares.

Objective: Combine both Positive & Negative number fill-ins [D] [3]. Add each square of numbers in any row, column, or diagonal to proof the same sum of 2056 for a solved Magic Square. Helpful hints below.

241		14											3		256
	226				27					22				239	
48		211											222		33
			196			58			55			205			
	79			181							188			66	
					166		89	88	171						
				108		151			154		101				
			125				136	137				116			
			141				120	121				132			
				156		103			106		149				
					86		169	168		91					
	191			69							76			178	
			52			202			199			61			
224		35											46		209
	18				235					230				31	
1		254											243		16

Magic Square
Number Fill-ins [D] [3]
Mathematical Formula 16 X 16 + 1 X 8 = 2056

SHELSEAS MAGIC SQUARE
Magical Stars 2056

Positive [D] Start with the bottom row, left to right in each row, going up. Begin with the number 1, count each square, only fill-in shaded squares. Diagonals are shown for the same sum of 2056. Helpful hints below.

241															256
	226													239	
		211											222		
			196									205			
				181							188				
					166					171					
						151			154						
							136	137							
							120	121							
						103			106						
					86					91					
				69							76				
			52									61			
		35											46		
	18													31	
1															16

Negative [D] Start with the top row, right to left in each row, going down. Begin with the number 5, count each square, only fill-in shaded squares. Helpful hints below.

				12							5				
					27					22					
						42			39						
							57	56							
80															65
	95													82	
		110											99		
			125									116			
			141									132			
		158											147		
	175													162	
192															177
						201	200								
					218				215						
				235						230					
				252							245				

Magic Square - Number Fill-ins [D] [4] - Mathematical Formula 16 X 16 + 1 X 8 = 2056

SHELSEAS MAGIC SQUARE
Magical Stars 2056

Positive [D] Start with the bottom row, left to right in each row, going up. Begin with the number 1, count each square, only fill-in same shaded squares. Diagonals are shown for the same sum of 2056.

Negative [D] Start with the top row, right to left in each row, going down. Begin with the number 5, count each square, only fill-in same shaded squares.

Objective: Combine both Positive & Negative number fill-ins [D] [4]. Add each square of numbers in any row, column, or diagonal to proof the same sum of 2056 for a solved Magic Square. Helpful hints below.

241				12							5				256
	226			27				22						239	
		211			42			39					222		
			196			57	56				205				
80				181							188				65
	95				166				171					82	
		110				151			154			99			
			125			136	137				116				
			141			120	121				132				
		158				103			106			147			
	175				86				91					162	
192				69							76				177
		52				201	200				61				
		35			218			215					46		
	18			235				230						31	
1				252							245				16

Magic Square
Number Fill-ins [D] [4]
Mathematical Formula 16 X 16 + 1 X 8 = 2056

SHELSEAS MAGIC SQUARE
Magical Stars 2056

Positive [D] Start with the bottom row, left to right in each row, going up. Begin with the number 1, count each square, only fill-in shaded squares. Diagonals are shown for the same sum of 2056. Helpful hints below.

1	2	3	4	5	6	7	8	9	10	11	12	13	14	15	16
241															256
	226													239	
		211											222		
			196									205			
				181							188				
					166					171					
						151			154						
							136	137							
							120	121							
						103			106						
					86					91					
				69							76				
			52									61			
		35											46		
	18													31	
1															16

Negative [D] Start with the top row, right to left in each row, going down. Begin with the number 4, count each square, only fill-in shaded squares. Helpful hints below.

1	2	3	4	5	6	7	8	9	10	11	12	13	14	15	16
			13									4			
				28							21				
					43					38					
						58			55						
							73	72							
96															81
	111													98	
		126											115		
		142											131		
	159													146	
176															161
							185	184							
						202			199						
					219					214					
				236							229				
			253									244			

Magic Square - Number Fill-ins [D] [5] - Mathematical Formula 16 X 16 + 1 X 8 = 2056

SHELSEAS MAGIC SQUARE
Magical Stars 2056

Positive [D] Start with the bottom row, left to right in each row, going up. Begin with the number 1, count each square, only fill-in same shaded squares. Diagonals are shown for the same sum of 2056.

Negative [D] Start with the top row, right to left in each row, going down. Begin with the number 4, count each square, only fill-in same shaded squares.

Objective: Combine both Positive & Negative number fill-ins [D] [5]. Add each square of numbers in any row, column, or diagonal to proof the same sum of 2056 for a solved Magic Square. Helpful hints below.

241			13									4			256
	226			28							21			239	
		211			43					38			222		
			196			58			55			205			
				181			73	72			188				
96					166					171					81
	111					151			154					98	
		126					136	137					115		
		142					120	121					131		
	159					103			106					146	
176					86					91					161
				69			185	184			76				
			52			202			199			61			
		35			219					214			46		
	18			236							229			31	
1			253									244			16

Magic Square
Number Fill-ins [D] [5]
Mathematical Formula 16 X 16 + 1 X 8 = 2056

Solved

[D]

Magic Square
Number Fill-ins [D]
Mathematical Formula 16 X 16 + 1 X 8 = 2056

SHELSEAS MAGIC SQUARE
Magical Stars 2056

Positive [D] Diagonals are shown for the same sum of 2056

241	242					247	248	249	250					255	256
225	226	227				232	233						238	239	240
	210	211	212	213						220	221	222	223		
		195	196	197	198					203	204	205	206		
		179	180	181	182					187	188	189	190		
			164	165	166	167			170	171	172	173			
145					150	151	152	153	154	155					160
129	130					135	136	137	138					143	144
113	114					119	120	121	122					127	128
97					102	103	104	105	106	107					112
			84	85	86	87			90	91	92	93			
		67	68	69	70					75	76	77	78		
		51	52	53	54					59	60	61	62		
	34	35	36	37						44	45	46	47		
17	18	19					24	25					30	31	32
1	2					7	8	9	10					15	16

Negative [D]

		14	13	12	11					6	5	4	3		
		29	28	27	26			23	22	21	20				
48					43	42	41	40	39	38					33
64	63					58	57	56	55					50	49
80	79					74	73	72	71					66	65
96	95	94					89	88					83	82	81
	111	110	109	108						101	100	99	98		
		126	125	124	123					118	117	116	115		
		142	141	140	139					134	133	132	131		
	159	158	157	156						149	148	147	146		
176	175	174					169	168					163	162	161
192	191				186	185	184	183						178	177
208	207				202	201	200	199						194	193
224					219	218	217	216	215	214					209
		237	236	235	234			231	230	229	228				
		254	253	252	251					246	245	244	243		

Magic Square
Number Fill-ins [D] [1]
Mathematical Formula 16 X 16 + 1 X 8 = 2056

SHELSEAS MAGIC SQUARE
Magical Stars 2056

Combination of both Positive & Negative number fill-ins [D] [1] is shown for a solved Magic Square. Objective: Add each square of numbers in any row, column, or diagonal to proof the same sum of 2056.

241	242	14	13	12	11	247	248	249	250	6	5	4	3	255	256
225	226	227	29	28	27	26	232	233	23	22	21	20	238	239	240
48	210	211	212	213	43	42	41	40	39	38	220	221	222	223	33
64	63	195	196	197	198	58	57	56	55	203	204	205	206	50	49
80	79	179	180	181	182	74	73	72	71	187	188	189	190	66	65
96	95	94	164	165	166	167	89	88	170	171	172	173	83	82	81
145	111	110	109	108	150	151	152	153	154	155	101	100	99	98	160
129	130	126	125	124	123	135	136	137	138	118	117	116	115	143	144
113	114	142	141	140	139	119	120	121	122	134	133	132	131	127	128
97	159	158	157	156	102	103	104	105	106	107	149	148	147	146	112
176	175	174	84	85	86	87	169	168	90	91	92	93	163	162	161
192	191	67	68	69	70	186	185	184	183	75	76	77	78	178	177
208	207	51	52	53	54	202	201	200	199	59	60	61	62	194	193
224	34	35	36	37	219	218	217	216	215	214	44	45	46	47	209
17	18	19	237	236	235	234	24	25	231	230	229	228	30	31	32
1	2	254	253	252	251	7	8	9	10	246	245	244	243	15	16

2056

Magic Square
Number Fill-ins [D] [1]
Mathematical Formula 16 X 16 + 1 X 8 = 2056

SHELSEAS MAGIC SQUARE
Magical Stars 2056

Positive [D] Diagonals are shown for the same sum of 2056

241	242	243	244									253	254	255	256
225	226	227		229							236		238	239	240
209		211	212		214					219		221	222		224
193			196	197		199			202		204	205			208
	178			181	182		184	185		187	188			191	
	162				166	167	168	169	170	171				175	
		147	148			151	152	153	154			157	158		
				133	134	135	136	137	138	139	140				
				117	118	119	120	121	122	123	124				
		99	100			103	104	105	106			109	110		
	82				86	87	88	89	90	91				95	
	66			69	70		72	73		75	76			79	
49		52	53		55			58		60	61				64
33		35	36		38				43		45	46			48
17	18	19		21							28		30	31	32
1	2	3	4									13	14	15	16

Negative [D]

				12	11	10	9	8	7	6	5				
			29		27	26	25	24	23	22		20			
	47			44		42	41	40	39		37			34	
	63	62			59		57	56		54			51	50	
80		78	77			74			71			68	67		65
96		94	93	92							85	84	83		81
112	111			108	107					102	101			98	97
128	127	126	125									116	115	114	113
144	143	142	141									132	131	130	129
160	159			156	155					150	149			146	145
176		174	173	172							165	164	163		161
192		190	189			186			183			180	179		177
	207	206			203		201	200		198			195	194	
	223			220		218	217	216	215		213			210	
			237		235	234	233	232	231	230		228			
				252	251	250	249	248	247	246	245				

Magic Square
Number Fill-ins [D] [2]
Mathematical Formula 16 X 16 + 1 X 8 = 2056

SHELSEAS MAGIC SQUARE
Magical Stars 2056

Combination of both Positive & Negative number fill-ins [D] [2] is shown for a solved Magic Square. Objective: Add each square of numbers in any row, column, or diagonal to proof the same sum of 2056.

241	242	243	244	12	11	10	9	8	7	6	5	253	254	255	256
225	226	227	29	229	27	26	25	24	23	22	236	20	238	239	240
209	47	211	212	44	214	42	41	40	39	219	37	221	222	34	224
193	63	62	196	197	59	199	57	56	202	54	204	205	51	50	208
80	178	78	77	181	182	74	184	185	71	187	188	68	67	191	65
96	162	94	93	92	166	167	168	169	170	171	85	84	83	175	81
112	111	147	148	108	107	151	152	153	154	102	101	157	158	98	97
128	127	126	125	133	134	135	136	137	138	139	140	116	115	114	113
144	143	142	141	117	118	119	120	121	122	123	124	132	131	130	129
160	159	99	100	156	155	103	104	105	106	150	149	109	110	146	145
176	82	174	173	172	86	87	88	89	90	91	165	164	163	95	161
192	66	190	189	69	70	186	72	73	183	75	76	180	179	79	177
49	207	206	52	53	203	55	201	200	58	198	60	61	195	194	64
33	223	35	36	220	38	218	217	216	215	43	213	45	46	210	48
17	18	19	237	21	235	234	233	232	231	230	28	228	30	31	32
1	2	3	4	252	251	250	249	248	247	246	245	13	14	15	16

2056

Magic Square
Number Fill-ins [D] [2]
Mathematical Formula 16 X 16 + 1 X 8 = 2056

SHELSEAS MAGIC SQUARE
Magical Stars 2056

Positive [D] Diagonals are shown for the same sum of 2056

241	242					247	248	249	250					255	256
225	226	227					232	233					238	239	240
	210	211	212	213							220	221	222	223	
		195	196	197	198					203	204	205	206		
			180	181	182	183			186	187	188	189			
			164	165	166	167			170	171	172	173			
145					150	151	152	153	154	155					160
129	130	131					136	137					142	143	144
113	114	115					120	121					126	127	128
97					102	103	104	105	106	107					112
			84	85	86	87			90	91	92	93			
			68	69	70	71			74	75	76	77			
		51	52	53	54					59	60	61	62		
	34	35	36	37							44	45	46	47	
17	18	19					24	25					30	31	32
1	2					7	8	9	10					15	16

Negative [D]

		14	13	12	11				6	5	4	3			
			29	28	27	26		23	22	21	20				
48				43	42	41	40	39	38						33
64	63			58	57	56	55							50	49
80	79	78				73	72						67	66	65
96	95	94				89	88						83	82	81
	111	110	109	108						101	100	99	98		
		125	124	123	122			119	118	117	116				
		141	140	139	138			135	134	133	132				
	159	158	157	156						149	148	147	146		
176	175	174				169	168						163	162	161
192	191	190				185	184						179	178	177
208	207			202	201	200	199							194	193
224				219	218	217	216	215	214						209
			237	236	235	234		231	230	229	228				
		254	253	252	251			246	245	244	243				

Magic Square
Number Fill-ins [D] [3]
Mathematical Formula 16 X 16 + 1 X 8 = 2056

SHELSEAS MAGIC SQUARE
Magical Stars 2056

Combination of both Positive & Negative number fill-ins [D] [3] is shown for a solved Magic Square. Objective: Add each square of numbers in any row, column, or diagonal to proof the same sum of 2056.

241	242	14	13	12	11	247	248	249	250	6	5	4	3	255	256
225	226	227	29	28	27	26	232	233	23	22	21	20	238	239	240
48	210	211	212	213	43	42	41	40	39	38	220	221	222	223	33
64	63	195	196	197	198	58	57	56	55	203	204	205	206	50	49
80	79	78	180	181	182	183	73	72	186	187	188	189	67	66	65
96	95	94	164	165	166	167	89	88	170	171	172	173	83	82	81
145	111	110	109	108	150	151	152	153	154	155	101	100	99	98	160
129	130	131	125	124	123	122	136	137	119	118	117	116	142	143	144
113	114	115	141	140	139	138	120	121	135	134	133	132	126	127	128
97	159	158	157	156	102	103	104	105	106	107	149	148	147	146	112
176	175	174	84	85	86	87	169	168	90	91	92	93	163	162	161
192	191	190	68	69	70	71	185	184	74	75	76	77	179	178	177
208	207	51	52	53	54	202	201	200	199	59	60	61	62	194	193
224	34	35	36	37	219	218	217	216	215	214	44	45	46	47	209
17	18	19	237	236	235	234	24	25	231	230	229	228	30	31	32
1	2	254	253	252	251	7	8	9	10	246	245	244	243	15	16

2056

Magic Square
Number Fill-ins [D] [3]
Mathematical Formula 16 X 16 + 1 X 8 = 2056

SHELSEAS MAGIC SQUARE
Magical Stars 2056

Positive [D] Diagonals are shown for the same sum of 2056

241	242	243	244									253	254	255	256
225	226	227				232	233						238	239	240
209	210	211	212									221	222	223	224
193		195	196	197							204	205	206		208
			180	181	182	183			186	187	188	189			
			165	166	167	168	169	170	171	172					
			149	150	151	152	153	154	155	156					
	130			134	135	136	137	138	139				143		
	114			118	119	120	121	122	123				127		
			101	102	103	104	105	106	107	108					
			85	86	87	88	89	90	91	92					
			68	69	70	71			74	75	76	77			
49		51	52	53							60	61	62		64
33	34	35	36									45	46	47	48
17	18	19				24	25						30	31	32
1	2	3	4									13	14	15	16

Negative [D]

				12	11	10	9	8	7	6	5				
			29	28	27	26			23	22	21	20			
				44	43	42	41	40	39	38	37				
	63			59	58	57	56	55	54					50	
80	79	78				73	72						67	66	65
96	95	94	93									84	83	82	81
112	111	110	109									100	99	98	97
128		126	125	124							117	116	115		113
144		142	141	140							133	132	131		129
160	159	158	157									148	147	146	145
176	175	174	173									164	163	162	161
192	191	190					185	184					179	178	177
	207			203	202	201	200	199	198				194		
				220	219	218	217	216	215	214	213				
			237	236	235	234			231	230	229	228			
				252	251	250	249	248	247	246	245				

Magic Square
Number Fill-ins [D] [4]
Mathematical Formula 16 X 16 + 1 X 8 = 2056

SHELSEAS MAGIC SQUARE
Magical Stars 2056

Combination of both Positive & Negative number fill-ins [D] [4] is shown for a solved Magic Square. Objective: Add each square of numbers in any row, column, or diagonal to proof the same sum of 2056.

241	242	243	244	12	11	10	9	8	7	6	5	253	254	255	256
225	226	227	29	28	27	26	232	233	23	22	21	20	238	239	240
209	210	211	212	44	43	42	41	40	39	38	37	221	222	223	224
193	63	195	196	197	59	58	57	56	55	54	204	205	206	50	208
80	79	78	180	181	182	183	73	72	186	187	188	189	67	66	65
96	95	94	93	165	166	167	168	169	170	171	172	84	83	82	81
112	111	110	109	149	150	151	152	153	154	155	156	100	99	98	97
128	130	126	125	124	134	135	136	137	138	139	117	116	115	143	113
144	114	142	141	140	118	119	120	121	122	123	133	132	131	127	129
160	159	158	157	101	102	103	104	105	106	107	108	148	147	146	145
176	175	174	173	85	86	87	88	89	90	91	92	164	163	162	161
192	191	190	68	69	70	71	185	184	74	75	76	77	179	178	177
49	207	51	52	53	203	202	201	200	199	198	60	61	62	194	64
33	34	35	36	220	219	218	217	216	215	214	213	45	46	47	48
17	18	19	237	236	235	234	24	25	231	230	229	228	30	31	32
1	2	3	4	252	251	250	249	248	247	246	245	13	14	15	16

2056

Magic Square
Number Fill-ins [D] [4]
Mathematical Formula 16 X 16 + 1 X 8 = 2056

SHELSEAS MAGIC SQUARE
Magical Stars 2056

Positive [D] Diagonals are shown for the same sum of 2056

241	242	243					248	249					254	255	256
225	226	227					232	233					238	239	240
209	210	211	212									221	222	223	224
		195	196	197	198				203	204	205	206			
			180	181	182	183			186	187	188	189			
			164	165	166	167			170	171	172	173			
			149	150	151	152	153	154	155	156					
129	130				135	136	137	138						143	144
113	114				119	120	121	122						127	128
			101	102	103	104	105	106	107	108					
			84	85	86	87			90	91	92	93			
			68	69	70	71			74	75	76	77			
		51	52	53	54				59	60	61	62			
33	34	35	36									45	46	47	48
17	18	19					24	25					30	31	32
1	2	3					8	9					14	15	16

Negative [D]

			13	12	11	10			7	6	5	4			
			29	28	27	26			23	22	21	20			
			44	43	42	41	40	39	38	37					
64	63				58	57	56	55						50	49
80	79	78					73	72					67	66	65
96	95	94					89	88					83	82	81
112	111	110	109									100	99	98	97
		126	125	124	123				118	117	116	115			
		142	141	140	139				134	133	132	131			
160	159	158	157									148	147	146	145
176	175	174					169	168					163	162	161
192	191	190					185	184					179	178	177
208	207				202	201	200	199						194	193
			220	219	218	217	216	215	214	213					
			237	236	235	234			231	230	229	228			
			253	252	251	250			247	246	245	244			

Magic Square
Number Fill-ins [D] [5]
Mathematical Formula 16 X 16 + 1 X 8 = 2056

SHELSEAS MAGIC SQUARE
Magical Stars 2056

Combination of both Positive & Negative number fill-ins [D] [5] is shown for a solved
Magic Square. Objective: Add each square of numbers in any row, column, or diagonal
to proof the same sum of 2056.

241	242	243	13	12	11	10	248	249	7	6	5	4	254	255	256
225	226	227	29	28	27	26	232	233	23	22	21	20	238	239	240
209	210	211	212	44	43	42	41	40	39	38	37	221	222	223	224
64	63	195	196	197	198	58	57	56	55	203	204	205	206	50	49
80	79	78	180	181	182	183	73	72	186	187	188	189	67	66	65
96	95	94	164	165	166	167	89	88	170	171	172	173	83	82	81
112	111	110	109	149	150	151	152	153	154	155	156	100	99	98	97
129	130	126	125	124	123	135	136	137	138	118	117	116	115	143	144
113	114	142	141	140	139	119	120	121	122	134	133	132	131	127	128
160	159	158	157	101	102	103	104	105	106	107	108	148	147	146	145
176	175	174	84	85	86	87	169	168	90	91	92	93	163	162	161
192	191	190	68	69	70	71	185	184	74	75	76	77	179	178	177
208	207	51	52	53	54	202	201	200	199	59	60	61	62	194	193
33	34	35	36	220	219	218	217	216	215	214	213	45	46	47	48
17	18	19	237	236	235	234	24	25	231	230	229	228	30	31	32
1	2	3	253	252	251	250	8	9	247	246	245	244	14	15	16

2056

Magic Square
Number Fill-ins [D] [5]
Mathematical Formula 16 X 16 + 1 X 8 = 2056

www.ingramcontent.com/pod-product-compliance
Lightning Source LLC
Chambersburg PA
CBHW080828180526
45168CB00006B/2612